ADVANCED DESIGN TOPICS FOR THE ENGINEERING OF AGRICULTURAL AND MECHANICAL EQUIPMENT

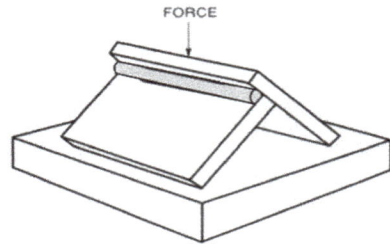

FORCE

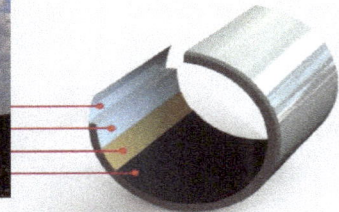

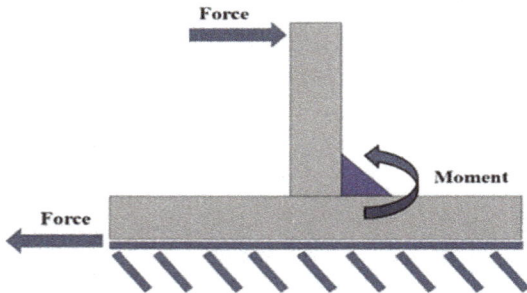

Force

Force

Moment

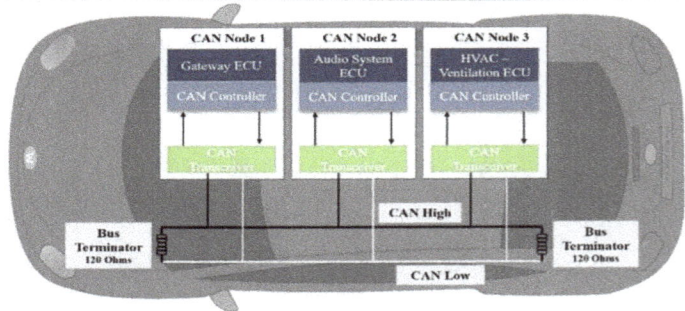

CAN Node 1
Gateway ECU
CAN Controller
CAN Transceiver

CAN Node 2
Audio System ECU
CAN Controller
CAN Transceiver

CAN Node 3
HVAC – Ventilation ECU
CAN Controller
CAN Transceiver

Bus Terminator 120 Ohms

CAN High

CAN Low

Bus Terminator 120 Ohms

ADVANCED DESIGN TOPICS FOR THE ENGINEERING OF AGRICULTURAL AND MECHANICAL EQUIPMENT

ROBERT M. STWALLEY III

ROGER TORMOEHLEN

PURDUE UNIVERSITY PRESS

WEST LAFAYETTE, INDIANA

Cataloging-in-Publication Data on file at the Library Congress.

978-1-62671-300-0 (paperback)

978-1-62671-301-7 (epdf)

Cover images: Industrial machinery springs: okeyphotos/iStock via Getty Images Plus; close-up of a used manure injector: RuudMorijn/iStock via Getty Images Plus; raking machine for haying with an old tractor: rudolfgeiger/iStock via Getty Images Plus; rear-view of modern agricultural tractor. Hydraulic hitch. Hydraulic lifting frame. Rear mechanism for attaching trailed equipment: Vladimir Zapletin/iStock via Getty Images Plus; harvest work. A modern combine harvester working a wheat field: fotokostic/iStock via Getty Images Plus.

The frontispiece page features images from the original articles. Case IH (CNH Industrial, N.V.) is acknowledged and thanked for their permission to use the photograph of the Stieger Quadtrac® Tractor. Bushing MFG is acknowledged and thanked for their permission to use the composite bearing drawing. The remaining drawings were all created by Robert M. Stwalley IV, who is acknowledged for his artwork and has granted his permission for its use. Permission for the use of all other photographs, drawings, and artwork within the original articles and subsequent republications was granted at the time of original publication by the creators and copyright holders.

For faculty of Purdue University Agricultural Engineering during the 1970s and 1980s. They were an incredible group of professors to learn from, and their fingerprints are all over everything that we, as the current instructors in Agricultural and Biological Engineering at Purdue, do today!

CONTENTS

Preface   xi

Acknowledgments   xiii

Introduction   1

Chapter 1. Mobile Electronic Systems: CAN-BUS   3
  1. Introduction and Background   5
  2. History   5
    2.1 CAN-BUS Development   5
    2.2 Early Applications of CAN Technology   7
  3. Technology Fundamentals   9
    3.1. CAN Utilization—Messaging Basics   9
  4. CAN-BUS Standards Development   13
  5. Implementation   16
    5.1 CAN-BUS by Industry Application   16
    5.2 Alternative Connectivity and Networking to CAN-BUS   18
    5.3 Prospective Areas for CAN Technology Inclusion   21
  6. Conclusions   24
  7. Acknowledgment   25
  8. Conflict of Interest   25
  9. CAN-BUS Questions   25
  10. References   26

Chapter 2. Soil Compaction   30
  1. Introduction   32
  2. The Relevance of Soil Compaction and Its Effects on Sustainability   32
    2.1. Environmental Impact of Soil Compaction   33
    2.2. Effects on Harvest Quality and Farmlands   34
    2.3. Social and Economic Impact   37
  3. The Causes of Soil Compaction   38
    3.1. Operation of Equipment with Heavy Axle Loads   38
    3.2. Operation during Non-Optimal Soil Conditions   39
    3.3. Livestock Grazing   40
    3.4. Other Factors   40
  4. Off-Road Vehicle Designs for Soil Compaction Management   41
    4.1. Tracks   41

4.2. Low-Inflation Tires   46

4.3. Two- versus Four-Wheel Drives   50

4.4. Sensors, Actuators, and Special Applications   52

5. Soil Compaction Management in Agriculture   55

5.1. Tillage Equipment and Practices   55

5.2. Tool Design   56

5.2.1. Structural Loading   59

5.2.2. Soil-Loosening Equipment   59

5.3. Controlled Traffic Farming   60

5.3.1. Low-Till   61

5.3.2. No-Till   62

5.3.3. Dedicated Tramline Equipment   62

5.3.4. Tillage Timing   63

5.4. Cover cropping and Crop Rotation   64

5.4.1. Cover cropping versus Crop Rotation   64

5.4.2. Cover crop Selection   66

6. Conclusion   66

7. Acknowledgments   67

8. Conflict of Interest   67

9. Soil Compaction Questions   68

10. References   68

Chapter 3. Composite Bearings   71

1. Introduction   73

2. Literature Review   75

3. Traditional Metallic Bearings   76

3.1. Plain Bearings   76

3.2. Antifriction Bearings   76

3.2.1. Ball Bearings   77

3.2.2. Roller Bearings   78

4. Composite Bearings   79

4.1. Materials   79

4.2. Lubrication and Friction   80

4.3. Strength and Loads   80

4.4. Thermal Properties   82

5. Composite Bearings in Heavy Off-Road Equipment   85

5.1. Reduction of Maintenance   86

5.2. Cost-Effectiveness   86

5.3. Resistance to Harsh Environments   87

5.4. Case Studies   88

5.4.1. Livestock Feed Mixer   88

5.4.2. Reduced Maintenance in Bulldozer Design   88

5.4.3 Composite Bushings in Earthmoving Equipment   88

6. Testing and Validation   89
7. Design Considerations   90
   7.4 Design Guides   90
   7.5 Basic Design and Manufacturing   91
   7.6 Product Availability and Procurement   92
8. Conclusion   93
9. Recommendations   94
10. Acknowledgments   95
11. Composite Bearing Questions   95
12. References   95

Chapter 4. Field Welding Practices   **98**
1. Introduction   99
2. Identification of Engineering Challenges in Repairs   101
   2.1. Repair Preparations   101
   2.2. Material Analysis   102
3. Welding Fundamentals and Equipment   105
   3.1. Arc Welding   105
   3.2. MIG Welding   108
   3.3. TIG Welding   108
   3.4. Welding Positions   109
      3.4.1. Flat Position   109
      3.4.2. Horizontal Position   110
      3.4.3. Vertical Position   110
      3.4.4. Overhead Position   110
   3.5. Weld Joint Types   113
      3.5.1. Fillet Welds   113
      3.5.2. Groove Welds   114
      3.5.3. Plug and Slot Welds   115
   3.6. Determining Weld Size   115
   3.7. Weld Strength   118
4. Challenges for Engineers and Technical Managers   120
   4.1. Weld Defects   120
   4.2. Repairing Weld Defects   122
   4.3. Weld Quality and Examination   122
   4.4. Filling Gaps and Repairing Cracked Components   123
5. Closure and Value of Welding Technical Knowledge   126
6. Acknowledgments   126
7. Conflict of Interest   126
8. Field Repair Welding Questions   127
9. References   127

Chapter 5. Theoretical Strength in Bending   **128**

1. Introduction   129
2. Literature Review   132
3. Experimental Procedures   132
4. Data Collection and Testing   139
5. Statistical Analysis   142
6. Summary of Findings   142
7. Theoretical Weld Bend Strength Questions   144
8. References   144

Chapter 6. Testing of Welds in Bending   **146**

1. Introduction   148
2. Background and Review   148
  2.1. Background   148
  2.2. Literature Review   150
  2.3. Objectives   153
3. Methodology and Materials   153
  3.1. Experimental Procedures   154
  3.2. Specimen Inspection Details   159
  3.3. Experimental Assumptions   162
4. Results and Discussion   162
  4.1. Bend Break Testing   164
  4.2. Break Test Root Fusion Inspections   164
  4.3. Statistical Analysis   165
    4.3.1. Normality of Data   166
    4.3.2. Data versus Cleaned Data   167
    4.3.3. Outliers from Defects and Impurities   169
  4.4. Failure Load versus Effective Weld Area   169
    4.4.1. Failure Load versus Effective Weld Area Analysis   173
    4.4.2. Theoretical Strength for Transverse Loading   174
  4.5. Allowable Stresses and Design Factors   175
    4.5.1. Allowable Stress in Tee-Joint Welding Tests   176
    4.5.2. Allowable Stress Experimental Statistics   176
  4.6. Discussion   176
5. Conclusion   177
6. Acknowledgments   179
7. Experimental Weld Bend Strength Questions   179
8. References   180

Conclusion   **183**

Combined References from the Included Works   185

About the Authors   197

PREFACE

One of the more challenging aspects of being a faculty member in a small engineering discipline is the difficulty in finding appropriate texts for student use in advanced courses. This compilation provides several peer-reviewed articles on advanced technical subjects of current and ongoing interest to field machine designers in Agricultural and Biological Engineering. These are generalized area topics that have applicability to agricultural and off-road machinery. Mobile computer systems, soil compaction, composite bearing use, and welding field repairs are covered in this initial installment of advanced design topics in our discipline. Each of these subjects is introduced with a preliminary essay, giving the students some focus on the key takeaways from each review. Reflective questions about the subject matter are provided following the essays for the instructor to utilize, which should demonstrate students' comprehension and mastery over the material. This collection is in no way intended to be a comprehensive set of all problems currently encountered in agricultural equipment design, but it is meant to be a supplemental text addressing the state-of-the-art in each subject area. This book may be used to provide a review of advanced topics in off-road equipment or as a comprehensive reference for each of these included design subjects.

The overall educational objectives are

- To be able to understand and utilize the data collection and analysis capabilities of mobile serial communication systems,
- To understand the impact of soil compaction and how propulsive and transport systems can minimize the impact of travel over production agricultural and silviculture soils,
- To become familiar with the capabilities and potential uses for composite bearings within off-road equipment, and
- To become familiar with the difficulties in performing welding repairs in the field and to be able to recommend appropriate procedures to adopt effective repairs.

ACKNOWLEDGMENTS

The efforts of the coauthor students from ABE 545, the Design of Off-Highway Machinery course at Purdue University—Ms. Hannah M. Boland, Ms. Morgan I. Burgett, Dr. Aaron J. Etienne, Mr. Michael M. Boland, Mr. Young U. Choi, Mr. Daniel G. Foley, Mr. Matthew S. Gobel, Mr. Nathan C. Sprague, Dr. Santiago Guevara-Ocana, Mr. Yury A. Kuleshov, Mr. Thomas Barnes, Mr. Brecken Beyer, Mr. Wyatt Griffey, Mr. Caleb Huffmeyer, Mr. Clayton Ness, Mr. Owen Nifong, and Mr. Tyler J. McPheron—are gratefully acknowledged for their overall work on these projects and the original publications. Dr. Carol S. Stwalley is thanked for her efforts in formatting this work for publication. The original publishers of the works contained in this volume—IntechOpen Limited, the American Society of Agricultural and Biological Engineers (ASABE), and Politeknik Negeri Lhokseumawe—are thanked for their efforts in the original dissemination of this work and for their permission to reprint these essays in this collection. The Purdue Department of Agricultural and Biological Engineering has also graciously helped provide backing for this project. An artificial intelligence–powered large language model software, ChatGPT, by Microsoft Incorporated (Redmond, Washington) was used to aid in the organization of some of the original essays, but all manuscript elements are original from the authors. Any errors are entirely the result of the authors. The mention of trade names and commercial products in this collection is solely for the purpose of providing specific technical information and does not imply recommendation or endorsement by Purdue University. The findings and conclusions in this publication are those of the authors and should not be construed to represent any official Purdue University determination or policy. Purdue University is an equal opportunity / equal access organization.

INTRODUCTION

The literal explosion in the size of off-road machinery, particularly in the agricultural arena, has eliminated numerous design challenges of the older generation of engineers and brought multiple new problems to the forefront for current developers. No one questions that diesel engine technology will continue to improve in mobile equipment and that the size of powerplants will increase. The current evidence would seem to show that in the past 65 years, tractor engine output sizes have grown by an order of magnitude, yet other issues that were unrecognized by previous designers have risen to potentially limit this growth and ultimately cap the size of propulsive agricultural tractors. As power and weight rise, so does the damage done to the soil by the passage of the equipment used to work the fields. Computers were rare academic and military devices in 1960, but today they are an everyday part of our lives, having a ubiquitous presence in our homes, offices, and vehicles. The incorporation of computational equipment into our agricultural machinery is ongoing, and its full potential cannot be predicted at this time. What is clear is that today's engineers and computer programmers have vastly greater amounts of information available to them than their predecessors had and that information will improve the efficiency of modern equipment to a level undreamed of by earlier designers. Modern technology is also available to those designers. Composite bearings can provide robust, maintenance-free motion joints for equipment operating under extremely challenging conditions. Improved best practices can even enhance the quality of maintenance and field repairs required in the post-manufacturing operational life of these products. These topics are included in this collection of subject reviews for the off-road vehicular industry for advanced students to reflect upon and utilize in their creative endeavors.

CHAPTER 1. MOBILE ELECTRONIC SYSTEMS: CAN-BUS

As the first quarter of the 21st century passes into the second quarter, controller area network binary unit systems (CAN-BUS) represent the predominate computational system on mobile equipment of all types. These serial computer systems, while somewhat primitive in certain aspects, provide a robust and expandable means to collect data, perform calculations, and control actuators on movable machinery. CAN-BUS systems can perform numerous functions on mobile machinery, including steering tractors, controlling earthmoving equipment blade depths, providing clean electrical power from emergency generators, trimming aircraft control surfaces, and modifying oceangoing ship ballast in heavy seas. CAN-BUS is very protocol-dominated, making new components compatible as "plug-and-play" devices. These systems are expandable, and they have built-in redundancies and robust error-checking algorithms that ensure error-free information transmission under demanding conditions. The world of mobile electronics is fast-paced and rapidly changing as consumers' demands from mobile equipment continue to become more complex and challenging. Engineers working with mobile equipment need a fundamental overview of the potential within CAN-BUS computer systems, and the following review, originally published in Technology in Agriculture, examines the history, capabilities, and future potential of CAN-BUS.

AN OVERVIEW OF CAN-BUS DEVELOPMENT, UTILIZATION, AND FUTURE POTENTIAL IN SERIAL NETWORK MESSAGING FOR OFF-ROAD MOBILE EQUIPMENT

HANNAH M. BOLAND, PURDUE UNIVERSITY DEPARTMENT
OF AGRICULTURAL AND BIOLOGICAL ENGINEERING

MORGAN I. BURGETT, PURDUE UNIVERSITY DEPARTMENT
OF AGRICULTURAL AND BIOLOGICAL ENGINEERING

AARON J. ETIENNE, PURDUE UNIVERSITY DEPARTMENT
OF AGRICULTURAL AND BIOLOGICAL ENGINEERING

ROBERT M. STWALLEY III, PURDUE UNIVERSITY DEPARTMENT
OF AGRICULTURAL AND BIOLOGICAL ENGINEERING

ABSTRACT

A controller area network (CAN) is a serial network information technology that facilitates the passing of information between electronic control units (ECUs), also known as nodes. Developed by BOSCH in 1986 to circumvent challenges in harness-connected systems and provide improved message handling in automobiles, the CAN interface allows broadcast communication between all connected ECUs within a vehicle's integrated electronic system through distributed control and decentralized measuring equipment. Since the early uses of CAN in car engine management, improvements in bit rate, bandwidth, and standardization protocols (such as ISO 11898 and SAE J1939) have led to CAN utilization in various industry applications, such as factory automation, aviation, off-highway vehicles, and telematics. Alternative wired and wireless technologies have been used to connect and network with CAN-BUS (Ethernet, Bluetooth, Wi-Fi, ZigBee, etc.), further expanding the diversity of applications in which the serial network is employed. In this chapter, the past, present, and prospective future developments of CAN technology, with focused attention on applications in the agricultural and off-road sectors, are broadly examined. CAN technology fundamentals, standards creation, modern day uses, and potential functionalities and challenges specific to CAN in the wake of precision agriculture and smart farming are discussed in detail.

KEYWORDS: CAN-BUS, serial network, agricultural sector, electronic control units

1. INTRODUCTION AND BACKGROUND

A controller area network (CAN) in a vehicle or a machine is analogous to the nervous system of a living organism. The nervous system of the body is a neuron-based network that collects signals from sensory receptors, passes chemical messages to and from the brain, responds to stimuli, and initiates actions. Expanding the analogy, sensors in a controller circuit are the equivalent of receptors, and an electronic control unit (ECU) can be visualized as a sensory neuron system dedicated to a specific function, bridging communication between the receptors and the central nervous system. CAN binary unit systems (BUS) create communication pathways between the electronic control units within a vehicle, allowing the transfer and interpretation of collected data. Prior to the invention of CAN-BUS, there was no efficient means of cross-communication between ECUs. CAN-BUS is efficient by relaying the most important messages first through a prioritization scheme of source ID-encoded messages using the BUS. This is an extremely robust arrangement, with a high ability to both detect signal errors and function when hardware is cross wired. This structure is fully distributed, which allows for a single access point for all the desirable information collected. The CAN-BUS is a relatively simple low-cost system that reduces the overall harness weight and amount of wiring needed in a vehicle, improving the integrity of transmitted data in comparison to harness-connected electrical structures [1].

While the CAN-BUS has been an effective communication technology in many past and present applications, future utilization of the network system continues to be a subject of research and development. In agricultural uses, this tool aids in precision agricultural applications and in the realm of data communication within larger farm systems. Vehicle autonomy is another area in which the CAN-BUS may play an important role as an intercommunication system. Additionally, there is still significant untapped potential for integrating CAN-BUS messaging into both more off-road control systems and wireless technologies.

The purpose of this chapter is to familiarize the reader with the importance of CAN-BUS in commercial off-road vehicles, applications, and future potential usage. In order to fully understand the benefits of CAN-BUS, the origins of CAN-BUS and its subsequent applications will be summarized. A high-level analysis of CAN-BUS technology, standards, and communication protocols will be presented to better familiarize the reader with essential technological concepts. Current applications of CAN-BUS and a comparison with alternative electronic control systems will be provided. A final qualitative evaluation of CAN-BUS capabilities will allow for a deeper understanding of why it is the dominant technology in modern vehicles and what innovations may be needed to expand its breadth of application in the changing technological landscape of off-road equipment.

2. HISTORY

2.1 CAN-BUS DEVELOPMENT

CAN was developed in 1986 by BOSCH as a means to overcome the limitations in harness-connected control systems [2]. BOSCH'S goal was greater functionality in message communication in automobiles, which could be accomplished through distributed control. A distributed control system connects multiple specific instrumentation into a system network that facilitates the transmission of data and information, adapting to the needs of the automation control scheme used. The system combines individual, decentralized measuring control equipment into a main network node, creating an interconnected network capable of controlling a larger system [3]. In developing the CAN system, the control equipment corresponded to nodes (or ECUs), which were connected to a two-wire bus, completing the network connection. The system prevented message collisions, thereby preventing the loss of crucial information, a common issue with other existing technologies at the time.

While other technologies could achieve the goal of internode communication, they required complex wiring systems, with each ECU individually connected to other ECUs to provide a communication pathway [1]. The point-to-point wiring of all ECUs was unnecessarily complex and caused difficulties in data and message management. In CAN-BUS implementation, all the connections are made directly on the same area network. Through utilization of microcontrollers, the system complexity decreased dramatically, allowing for a reduction in wiring, a simplified manufacturing assembly process for connecting nodes, and an overall increased system performance. Due to the improved efficiencies and system simplicity that this technology offered, CAN-BUS became a viable alternative to the complex point-to-point wiring harnesses used at the time [4].

In 1987, both Intel and Philips developed the first CAN controller chips, the Intel 82526 and the Philips 82C200, respectively [2]. The first iteration of this technology was a chip that managed messages by assigned priorities. This allowed the more important messages to be received with significantly less delay. Notably, this first system included error detection, which would automatically disconnect faulty nodes while still allowing uninterrupted communication between working nodes [5]. The hierarchy system allowed for the most crucial information to be passed along first, making the system particularly useful in applications with high safety requirements [1].

In early CAN development, there were two hardware implementations that cover the bulk of installations: Basic CAN and Full CAN. Basic CAN utilized a single message buffer to receive and transmit messages. The standard CAN controller implemented a specified number of message buffers (usually around 16), wherein the programmed algorithm read the received messages and wrote messages to be transmitted [6]. In Basic CAN, the received message is passed through acceptance filtering, which then decides whether to process a message or ignore it. Software is used to control the acceptance filtering of a node in Basic CAN. Bit masks for message identifiers (IDs) make it possible to ignore certain messages by ignoring specific IDs in order to reduce the software load requirement at the individual nodes [7].

Compared to Basic CAN, Full CAN is a bit more complex. Every transmitted or received message is accompanied by 8 to 16 memory buffers in the Full CAN scheme. Hardware, rather than software, performs acceptance filtering in this system, reducing the overall software load significantly. Individual buffers are configured to accept messages with specific IDs, and unique buffers for individual messages allow more processing time for the messages that are received. The transmitted messages can then be better handled according to their priority levels. Data consistency is also improved through this one-on-one buffer-to-message configuration [7]. Unfortunately, Full CAN is limited in the number of frames that can be received and requires more computational chips at each node than Basic CAN. Early CAN controllers by Intel and Philips were constructed under the Basic CAN or Full CAN configurations, with Philips favoring the former and Intel the latter. Modern CAN controllers combine the frame handling and acceptance filtering strengths of both, so the distinction is no longer made between Basic CAN and Full CAN [2].

A major milestone in bringing CAN-BUS into industry was the development of the CAN-in-Automation (CiA) working group in 1992. The CiA is an international organization composed of manufacturers and users with the goal of creating developmental content based on members' interests and initiatives [2]. One year later, the International Organization for Standardization (ISO) published ISO 11898, which defined CAN communication protocols for the automotive industry. ISO is a nongovernmental organization, without corporate affiliations, composed of individual standards organizations from 165 nations. The ISO develops voluntary international standards and improves the world's trading potential by providing common standards across the globe [8]. The implementation of an ISO standard for CAN-BUS was an important step in bringing coherence and marketability to the serial network system.

As the bandwidth requirements of the automotive industry continued to increase, the CAN data link layer (which will be covered in later sections) needed to be updated. BOSCH began developing the CAN flexible data rate (FD) protocol in 2011, working in conjunction with carmakers and other CAN experts. This updated protocol surmounted two of the most restrictive early CAN limitations: the data transfer rate and payload. CAN FD allows for a bit rate (transmission speed) of up to 12 megabits per second (Mbps), 12 times faster than the previous maximum transmission limit. The data field message payload was expanded up to 64 bytes in length, an increase of eight times beyond the previous payload size restriction [2]. CAN FD incorporated a simple yet powerful ideology: when only one node is transmitting data, the bit rate can increase, as no nodes need to be synchronized. The nodes are then resynchronized following the data transmission and data integrity check, just prior to an acknowledgment of data acceptance [9]. By 2015, ISO 11898 had been updated to incorporate CAN FD, which has continued to be the standard CAN system in commercial implementations [2].

2.2 EARLY APPLICATIONS OF CAN TECHNOLOGY

CAN-BUS has played a major role in industry since its debut in 1987. In the mid-1990s, companies such as Infineon Technologies and Motorola began shipping large quantities of CAN-BUS controllers to European automotive manufacturers, marking the advent of CAN utilization in the automotive industry. In 1992, Mercedes-Benz was noted as the first manufacturer to implement the controller within its processes, when CAN-BUS was first incorporated in its high-end passenger cars for engine care management [2].

In 1995, BMW was next to implement CAN-BUS technology. BMW introduced a star topology network with five electronic control units in its 7 Series cars. Then, the company took the implementation even further and employed a second network for body electronics. This allowed two separate CAN-BUS networks to be associated through gateway connections. Following BMW's example, other manufacturers soon began implementing two separate systems in all their passenger cars. Today, many manufacturers have multiple CAN-BUS networks associated with their production vehicles [2]. An example of vehicular integration is presented in Figure 1.1.

In 1993, a European consortium led by BOSCH prototyped a network that would later become CANopen. This project was eventually passed to CiA for further development and maintenance. In 1995, CANopen was completely revised and became the most important standardized network in Europe within just a few years. The CANopen network protocol offers high configuration flexibility, which has allowed its installation in a multitude of applications. The networks were first used for internal machine communications, specifically in drives, but they have since been utilized in many other industries. In the United States, CANopen has been implemented for use in forklifts, letter sorting machines, and other network processes [2].

As mentioned in the previous section, introduction of CAN-BUS into the automotive world required the standardization of protocols and testing standards to ensure CAN system conformity. ISO 11898, the first international standard for CAN, was based on the BOSCH CAN specification 2.0, and it standardized the

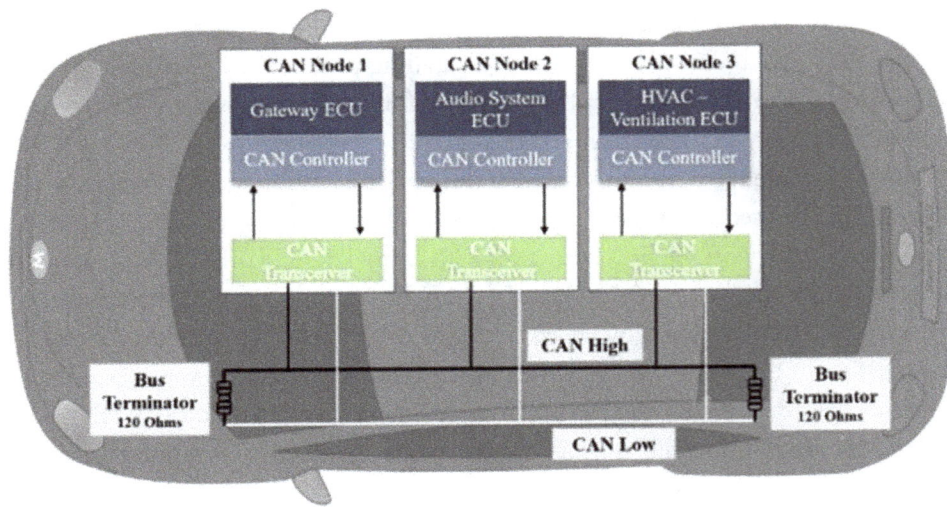

FIGURE 1.1. Illustration showing the multiple node connections to CAN-BUS in a modern vehicle.

high-speed physical layer for the system at the time [10]. As network technology continued to develop, allowing for different data transmission speeds and fault tolerances in the physical layers, new revisions to standards and interfaces for vehicle-specific applications were needed. This led to the development of SAE J1939 for heavy-duty vehicles and multiple other ISO standards (some will be covered in Section 4 in this chapter). Due to the rapidity of CAN modification and development in the early 1990s, no error-free complete standards or CAN specifications were available for CAN chip manufacturers. This led to the establishment of CAN conformance testing houses, where all CAN chips could be tested for compliance to the BOSCH CAN reference model using the testing plans outlined in ISO 16845 [2]. These steps were important in allowing the new technology to be widely applied in a variety of markets.

With regard to the marketing of CAN-BUS into the agricultural industry, in 2000 the German Mechanical Engineering Professional Society founded the Implementation Group of ISOBUS to promote the ISOBUS controller. The German Agricultural Society (Deutsche Landwirtschafts-Gesellschaft) assisted with the development of the first tests and a testing facility for ISOBUS compliance, which remains the primary test house for device compatibility. In 2009, several companies joined to form the Agricultural Industry Electronics Foundation, a nonprofit organization that further promoted the use of CAN-BUS controllers, especially the implementation of ISO 11783. Since then, there have been many plug-tests organized at various locations. The first plug-test for CAN-BUS in North America was hosted by the Nebraska Tractor Test Laboratory in 2010 [11].

This review of the development of CAN-BUS and its early applications illustrates some of the current and future directions for the technology. Besides the novel use of a distributed communication network, these development efforts have truly positioned CAN-BUS as the leading serial network system in off-road vehicles. The establishment of international societies and standards has been essential in this effort. The societies are dedicated to enforcing CAN standardization across the industry and to enhancing the functionality and quality of CAN technology through research and development. These organizations will likely continue to play an important role as CAN systems are utilized in new implementations going forward.

3. TECHNOLOGY FUNDAMENTALS

3.1 CAN UTILIZATION: MESSAGING BASICS

To gain a more complete grasp of how CAN ID messaging works and how different ECUs can interpret these messages, it is helpful to understand the overall structure of CAN messages from both a data and a hardware perspective. This section covers the physical architecture of the BUS, the different components of CAN messages, CAN error-handling, and a high-level breakdown of CAN layers and provides an overview on how CAN-BUS supports effective messaging channels.

The physical architecture or layer of a CAN includes two wires, CAN High (CAN-H) and CAN Low (CAN-L), which carry all CAN messages between ECUs and connect to BUS terminators at each end. The BUS terminators are powered and grounded, providing the necessary voltage to allow serial network operation. The most standard form of CAN wiring in modern systems is the twisted quad cabling configuration, in which a terminating bias circuit, with a power wire and a ground wire, is wound together with the CAN-H and CAN-L signal wires between the two terminators [12]. Both of the signal wires have set dominant and recessive voltages that correspond to the CAN system type (high speed or low speed). The system reads the voltage difference between the two wires as a bit value of "0" when the voltages are dominant, or a value of "1" when the voltages are recessive, creating the mechanism of sending binary messages through the system hardware [13].

A maximum of 30 ECUs can be attached to a single section of the BUS, and the overall number for ECUs connected to the network is limited to 254. The maximum number of available ECU addresses is limited to 256 because the maximum length of a data signal is 8 bytes. The 255 address is left null, and the 256 address indicates that a message be accepted by every ECU connected to the network [12]. Since CAN-BUS is a broadcast protocol, messages are not sent to specific nodes; rather, every ECU connected to the network receives every transmission from all other nodes on the same network. Various ECUs typically have filters on their receiving ends so that the local computer only accepts the messages that pertain to its operational needs [14]. This open communication between all connected nodes helps to improve the manufacturing process and implementation of the system, creating vehicle-wide interconnection. Since all the nodes are linked by subsystem functions, there are no redundant connections between any two specific ECUs.

As shown in Figure 1.2, a basic CAN message has eight key parts: (1) start of frame (SOF), (2) CAN identifier (CAN ID), (3) remote transmission request (RTR), (4) control, (5) data, (6) cyclic redundancy check (CRC), (7) acknowledgment (ACK), and (8) end of frame (EOF). It should be noted that the CAN frame consists of parts 2 and 5: the CAN ID and the data [12]. The SOF is a 1-bit "dominant zero" at the beginning of a CAN message that signals that an ECU is about to send a message. This alerts other ECUs connected to the CAN to "listen" for the message transmission. The CAN ID contains information on the message priority (lower values indicate higher priority) and the source address. The ID bit length varies by version of CAN, with CAN 2.0 being 11 bits and later versions relying on extended 29-bit IDs. The RTR is another 1-bit piece of the message indicating whether a node is sending data to or requesting data from a specific ECU. The control portion of a CAN message is 6 bits in length, 4 bits of which are the data length code, which denotes the size of the data message to be transmitted (0–8 bytes) [13]. The data segment of the CAN message makes up the bulk of information being communicated and contains all the CAN signals to be extracted and decoded for use by the receiving ECUs [5].

Standard CAN Data Frame Structure

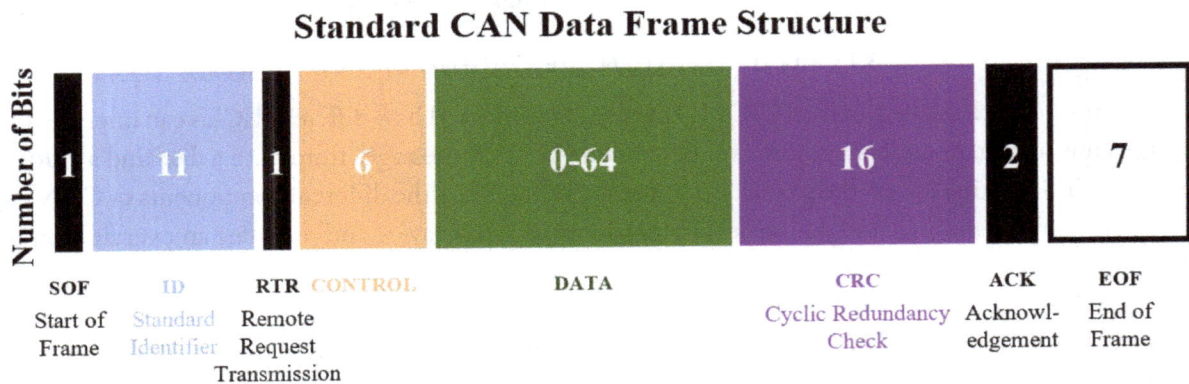

FIGURE 1.2. CAN-BUS message structure.

The four message parts prior to the data portion are all used to give the receiving ECUs adequate information on whether to receive the data being sent and what kind of data to expect. The last three parts of a CAN message are used to ensure that the data was transmitted successfully. The CRC is a 16-bit portion of the data that checks the data integrity, while the ACK is a 2-bit acknowledgment that the CRC found no issues with the data, allowing it to pass. Finally, the EOF is the 7-bit cap on a CAN message that signals the end of the transmission [13]. A breakdown of these eight parts highlights the strength of CAN messaging in that it provides both front-end and back-end contexts for the data being sent. Message types used in CAN-BUS include the data frame (a data transmission message), the error frame (a message that violates CAN formatting to signal an error in data transfer), the remote frame (a message to request data), and an overload frame (a message transmitted by an overloaded node to trigger delays) [5].

System robustness and error handling are the two major benefits of the CAN-BUS architecture. Error handling is the methodology of detecting flawed messages that come across the CAN-BUS, in which the original sender destroys a faulty message using an error frame, and then retransmits the correct message. All CAN controllers connected to the BUS listen for potential transmission errors whenever a new message is sent along the BUS [15]. When an error has been identified, the node that discovered the error will transmit an error flag throughout the system, halting all CAN-BUS traffic. The other connected nodes will each receive the error flag and transmit eight recessive bits, known as an error delimiter signal, to clear the BUS before taking appropriate action in response to the error. The most common response to an error flag is to discard the erroneous message and continue to transmit and receive other messages streaming on the BUS. This allows for what is known as fault tolerance, or the ability for the system to function around an error state [15]. An example of the error handling message structure is detailed in Figure 1.3.

Each node keeps a record of detected errors through two different registers. Errors that the ECU was responsible for sending are accounted for in the transmit error counter, while faults that it detected in other nodes' messages are logged in the receive error counter. Several protocols have been defined that govern how recorded errors increment or decrement the counters. When a transmitter detects a fault error in a message, it increments the register for the transmit errors at a faster rate than the receiving nodes increment their receive error registers, since the transmitter causes system faults in most cases. When a node's error counter

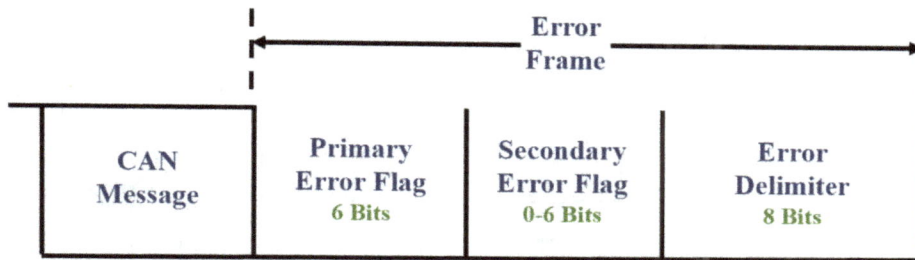

FIGURE 1.3. A sample of an error handling message structure.

exceeds a predetermined value, the ECU enters an error passive state, in which its error detection activities won't be broadcast on BUS traffic for other nodes to see. When the counter rises above a second higher preset value, it switches into a BUS-Off state, removing the ECU from participation in BUS traffic [15]. Through this process, CAN nodes can both detect faults and perform error confinement.

An Open Systems Interconnect (OSI) reference model is utilized by CAN-based network solutions. This same standard is applied across all modern communication technologies. This model is standardized in ISO/IEC 7498-1, which defines "a common basis for the coordination of standards development for the purpose of systems interconnection" [9]. The adapted CAN message model comprises three of the seven OSI layers: the first layer, the CAN physical layer; the second layer, the CAN data link layer; and the seventh layer, the CAN application layer. Typically, OSI layers 3 through 6 (network, transport, session, and presentation layers) are not explicitly implemented. It is common for the application layers in CAN to incorporate functions of network and transport layers to allow this adaption of the OSI model without sacrificing functionality [16].

Higher layer protocol functionality, which spans between the network and application layers, is an important factor in CAN network design. Network management, which includes the protocol for turning CAN nodes on and off, can be included in this functionality. Node supervision in event-driven networks is another common function in network management [17]. This supervision is required to detect nodes that are missing due to several possible fault conditions. Missing nodes could be caused from a BUS-Off state, a temporary power loss, or a permanent power loss. Application layers can search for missing nodes using one of two methods. For nodes that do not transmit messages periodically, a client/server service can be programmed so that a connected server sends a state message to the monitoring "client" after a consistent period, providing a "pulse." Any interruption to the pulse that exceeds a set time limit indicates an offline status in that node. However, if the node does transmit messages in a periodic fashion, this detection can be done implicitly [16]. An example of this time-out utilization in error reporting is given in Figure 1.4.

One of the most significant higher-layer protocol services in CAN is breaking up data for transmission and reassembling it on the receiving end. While this function is typically associated with the transport layer in OSI, in CAN this parsing of data is another role executed by the application layer. Examples of protocols that provide this service include CANopen, DeviceNet, and J1939-21 [17]. Device and network design have become simplified through the utilization of software routines that execute standardized higher-layer protocols. These protocols are typically implemented in software through protocol stacks. Standardized versions of these stacks are commonly available from a variety of manufacturers. Examples of these standardized protocol stacks

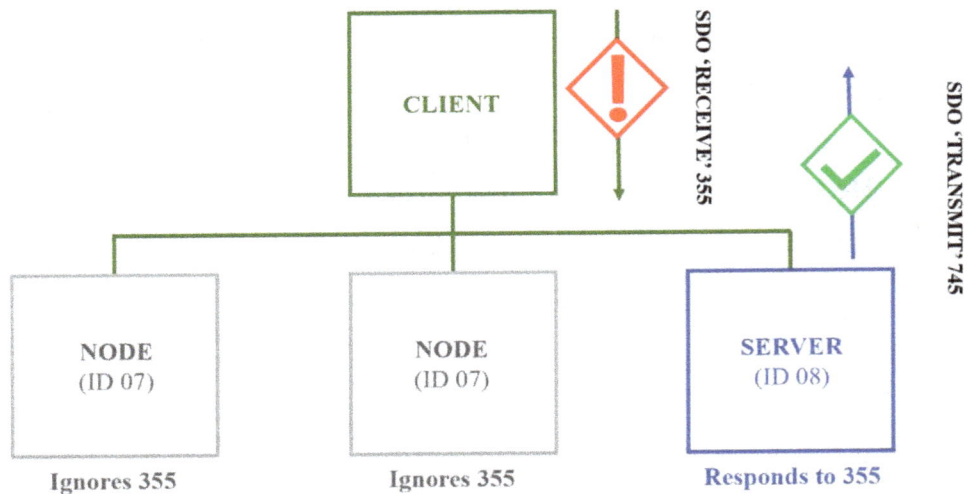

FIGURE 1.4. Implicit message time-out reporting utilizing CANopen.

include CAN Application Layer (CAL) from CiA, NMEA 2000 from the International Electrotechnical Commission, and CAN FD from CiA.

CAN-BUS, as an overarching protocol for vehicle system-to-system communication, helps the vehicle make informed decisions about component-level maintenance and control by maintaining an efficient communication pathway. To facilitate effective information flow, there are often multiple levels and separate systems of CAN that control specific regions and subsystems of the vehicle. This improves information handling capacity and also helps simplify the system into subsets that only contain the ECUs that need to communicate with each other. There is no reason, for example, for the ECU controlling in-cab climate control to know what is happening with the left rear tire pressure sensor. These controllers are divided onto different specialized networks, enhancing system efficiency.

In addition to separating networks into subsystems, there are also different types of CAN-BUS systems that allow for different speeds of communication. The high-speed CAN system uses the CAN-H and CAN-L wires described above and can communicate at speeds up to 1 Mbps. The ECUs that require this high communication speed are safety-critical systems, such as the engine electronic control unit, the brake controller, and the air pollution control systems [12]. These are wired in a linear serial bus configuration terminated by resistors. The other type of CAN-BUS commonly used is low-speed CAN, which can only reach communication speeds of up to 125 kbps. This is an eighth of the high-speed system rate and is appropriate for fault-tolerant or comfort systems such as cab climate control and interior lights. A star serial bus configuration may be used, where multiple CAN applications are terminated at nodes [4]. By splitting up the networks, there is a higher level of reliability for safety-critical systems to get their messages broadcast across the network. This can aid in the avoidance of accidents and in notifying a driver of an in-process component failure, such as the loss of engine oil pressure.

To further improve efficiency of CAN-BUS, every ECU on the network is also assigned an arbitration ID, or an identification number. This ID dictates which ECU is given priority in case there are conflicting messages or messages sent at the same time. This priority framework is a large part what makes CAN so efficient.

Important messages from the engine regarding fuel input, for example, are not delayed by a message from the oil pump that oil life has decreased by 1%. In having an established priority level of messages, the system can be sure that system-critical messages are broadcast and received across all interconnected ECUs. This system of broadcasting the highest-priority message has been a main contributor to the success of CAN-BUS technology and its dominance in the market.

While CANs are effective at communicating data between ECUs, they can also be utilized to record the operational metrics of a vehicle. Instead of directly measuring the data with precision instruments, approximate results can be calculated using the theoretical relationships between a specific metric and other parameters that are measured with internal sensors on the CAN. These internal sensors are commonly found in plug-and-play tools that are widely available on the market for onboard processing and diagnostics. They generally have low customizability, but they are very simple to install when compared with more specialized auxiliary sensing equipment [18]. While estimates from these embedded controllers are inexact, very accurate measurements can be obtained via this method by first calibrating the internal sensors with precision external sensors, as shown in Polcar, Čupera, and Kumbár's study on fuel consumption measurement [19]. This allows a reduction in both the number of sensors and the overall cost required within a vehicle's control system.

Through its methods of system interconnection and communication, CAN-BUS has revolutionized data collection and autonomy in virtually all markets, especially in the agricultural industry. By splitting up the various subsystems to create an efficient communication pathway between the multiple electronic control systems that need to communicate, CAN-BUS has become an invaluable addition to modern agricultural equipment and continues to advance the capability for onboard real-time data collection, providing farmers with sophisticated technologies for improving their operations.

4. CAN-BUS STANDARDS DEVELOPMENT

Thus far, this chapter has made references to CAN standards, such as ISO 11898 and SAE J1939, but has not given an explanation as to why there are different standards for different vehicle types. This section will discuss the purpose and need for developing such individual industry standards and also introduce some of the most important CAN standards in industry today, especially with respect to agricultural vehicles.

As previously mentioned, CANs function using a serial communication protocol, making it a useful pathway for passing digital data. However, without a standard for interpreting and forwarding the data, no useful information or actionable processes can be gleaned from it. Using the analogy of a telephone, CAN would be equivalent to the hardware and telephone lines used to connect the voices of two individuals, while the standard is the language used to make the communication meaningful [5]. Just as it is important that the individuals on opposite ends of the telephone line use the same language conventions to interpret each other's speech, the same is true with standard compatibility within a vehicle's system. Many components in a single vehicle are produced by different manufacturers, and standards allow the ECUs of these various modules to function and communicate on a common network.

The first standards were focused primarily on CAN usage in automobiles, as engine care management was the original target market for usage [2]. As off-road and heavy-duty vehicles carry out entirely different mission profiles from passenger cars, with respect to loads, implement usage, and speed, it was not possible to apply the same "language" for priority and layer management in these vehicles. This led to the evolution of application-specific standards for the vehicle manufacturing industry. To give some more context for what these standards entail, ISO 11898, SAE J1939, and ISO 11783 will be covered briefly.

ISO 11898 was released in 1993. It was initially divided into two parts, and a third part was added later. This standard covers the data link layer, the physical layer for high-speed medium attachment (HS-PMA), and the physical layer for a fault-tolerant, low-speed, medium-dependent interface. ISO 11898-1 gives the specifications for creating an interchange of data between the modules of the CAN data link layer [10]. ISO 11898-1 also specifies the two main format options, the Classical CAN frame format and the CAN Flexible Data Rate format, the latter of which was introduced in 2012. While Classical CAN supports a maximum bit rate and payload of 1 Mbps and 8 bytes per frame, respectively, the Flexible Data Rate frame format extends the allowance for both bit rate and payload beyond these original limits. The general architecture of CAN is also described in this ISO standard in terms of the OSI layers mentioned previously. The ISO standard contains specifications for both the logical link and medium-access control sublayers and the physical coding sublayer [6]. ISO 11898-2 gives the specifications for HS-PMA, which is a serial communication protocol that allows for real-time control of components in vehicle systems by multiplexing data for immediate use. The standard formalizes HS-PMAs with low-power mode and selective wake-up options [20]. ISO 11898-3 additionally covers the setup of a data exchange between the ECUs of a vehicle utilizing CAN [21].

SAE J1939 was developed by the Society of Automotive Engineers (SAE) in 1994 and establishes how nodes transmit data on the CAN-BUS in heavy-duty vehicles [22]. J1939 provides a common communication language across heavy equipment from different manufacturers, allowing a wide range of equipment to work with each other and enabling consistent data logging across heavy-duty and off-road vehicles. Although the first standards development papers on J1939 were drafted in 1994 (J1939-11, J1939-21, and J1939-31), it was six years before the initial top-level document was published. After this, CANs were officially included within the language of the standard. In 2001, J1939 replaced the older standards SAE J1708 and SAE J1587. J1939, along with its accompanying documents, has since become a wider industry standard and is currently utilized for applications across multiple industries, including agricultural machinery, construction equipment, forestry machines, maritime ships, mass transportation, material handling, and military applications [1].

There are several key characteristics that define SAE J1939. Its bit rate, or the speed at which messages travel across the BUS, was originally set at 250 kbps. More recently, the standard was updated to support a faster bit rate of 500 kbps, and the ID, or unique name of each message, was extended to 29 bits. The message ID segment, in addition to describing its data content and message priority and indicating the source address, is also used in J1939 to specify the destination on the network [23]. The primary differentiation in message composition from other CAN systems comes from the Parameter Group Number (PGN). This 18-bit PGN is a function-specific frame sandwiched between the first 9 and last 2 bits of a traditional 11-bit CAN ID, providing more detail regarding the message content [24]. The J1939 message parameters within data bytes are identified by Suspect Parameter Numbers (SPNs). SPNs correlate to specific PGNs, with their encoded data designated by bit start position, bit length, scale, offset, and units. These PGN-specific details are used to extract desired SPN data and convert it to meaningful physical values [25]. An illustration of the J1939 message structure is shown in Figure 1.5.

Development of ISO 11783, a CAN-based agricultural BUS by Landwirtschaft Bussysteme (LBS), began in the early 1990s with the German DIN 9684 standard. The first commercially successful Landwirtschaft

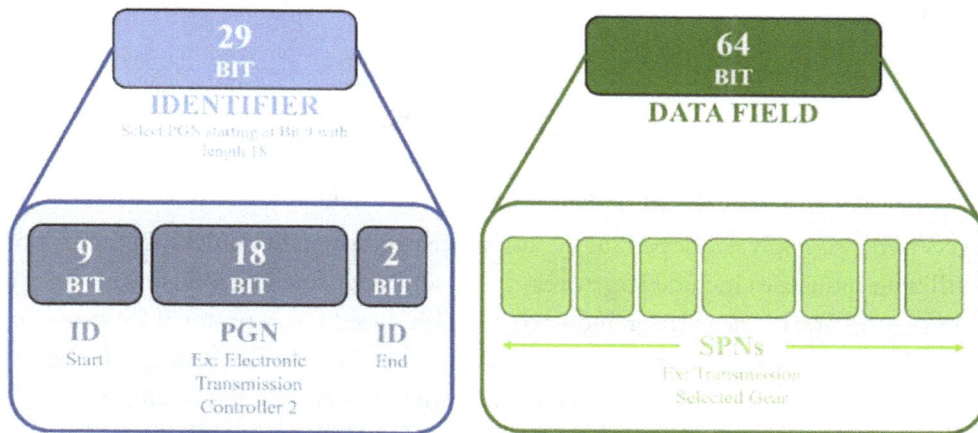

FIGURE 1.5. SAE J1939 message structure.

Bussysteme BUS combined the DIN 9684 virtual terminal (VT) concept with J1939 protocols and was internationally standardized as the ISO 11783 series [11]. The accompanying BUS detailed in this standard is commonly known as ISOBUS. This standard consists of 10 specific parts, including (1) the general standard for data communication, (2) the physical layer, (3) the data link layer, (4) the network layer, (5) network management, (6) the virtual terminal, (7) the implement messages applications layer, (8) power train messages, (9) the tractor ECU, and (10) the task controller and management computer interface [14]. The communication protocols define messaging between the tractor and implement electronic systems through CAN. These, combined with the serial data network, regulate the methodology of data transference between actuators, control elements, display units, information storage systems, and sensors, allowing the tractor to control an implement through the VT.

The VT is one of the most important features of the ISO 11783 standard, as it allows the operator to interface with the tractor and implements by both viewing real-time data and providing user inputs. The VT acts as a slave to individual ECUs, each of which secure terminal connectivity to display informational data and collect operator inputs according to their individual protocol. The operator can choose which operational data to display, while each connected ECU continues to operate as if the VT were dedicated solely to its specific function [14]. This pathway makes it possible for the operator to have greater control over the functions of an implement, such as sprayer nozzle flow, combine cylinder rotational speed, and cultivator attachment height, depending on input from implement sensors. This eliminates the need for a separate control box for the implement and provides a single terminal controlling all information flow to the operator [11]. The ISOBUS is based on CAN running at 250 kbps and uses twisted non-shielded quad cable and high-speed transceivers (the same as ISO 11898). A nine-pin connector on the tractor is the only required point of contact between it and an implement with an ISOBUS-compliant network cable.

This overview of CAN communication and standards has presented a cursory background of the technology fundamentals associated with the serial networking scheme as well as some brief mention regarding how it is implemented. The next section will go into greater depth on how CANs have been utilized in industry, the potential connection of CANs to other network technologies, and how CAN usage could be expanded in the future.

5. IMPLEMENTATION

5.1 CAN-BUS BY INDUSTRY APPLICATION

Although CANs were originally developed for the automotive industry, they quickly became popular in other areas. CAN-utilizing industries include large over-the-road trucks, forestry, industrial factory automation, aerospace, and many others. In the aviation industry, the high-speed CAN protocol ISO 11898 is widely utilized, along with ARINC 825, a protocol created specifically for the aviation industry. The effort to create a CAN-based standard for communication in aircraft was initiated by Airbus and Boeing and was advanced by the Airlines Electronic Engineering Committee through its CAN Technical Working Group [26]. Several design targets were set while developing this protocol, including CAN functionality as either a main or ancillary network, an allowance for local CAN network integration into the wider aircraft network, and interoperability and interchangeability of CAN connected line replaceable units. Other design mandates were to maintain flexible configuration options; establish a simple process for adding, deleting, or modifying BUS ECUs; and simplify systems' interconnection protocols [26].

CAN-BUS systems also play an important role in both modern factory automation processes and testing facilities. Since CAN design is based on distributed control principles, it has been effectively used in manufacturing facilities to connect the essential control systems dispersed throughout a plant. Through the use of human machine interfaces, operator inputs can be translated into instructions that a programmable logic controller dispatches onto the BUS, allowing the remote operation of equipment ranging from sensors to actuators. This process allows the testing of new input parameters prior to execution on specific equipment and is a viable option for increasing process safety [27]. Use of CAN on assembly lines as a quality check is also becoming more common and is especially important on a line manufacturing a customizable product. Certain specifications are programmed for each checkpoint of product assembly, which are then broadcast on the CAN between machines to provide quality validation for the operators throughout the manufacturing process. CAN-BUS is also a practical option for connecting security and environmental control systems across a facility due to both high bit rate and inexpensive installation. [27].

Returning to CAN use in the off-road vehicle market, virtually all modern agricultural machines incorporate CAN-BUS. Improved vehicle diagnostics, less complex design of electronic circuit controls, and advanced implement management are all benefits that CAN-BUS technology brings to the agricultural sector. CAN-BUS allow for high precision in machinery performance and logistics information. These metrics help to estimate operational cost and projected size in downstream operations. Specific measurement of other metrics, including fuel consumption, engine load, and average operating speed, can also help supply chain managers maximize field and transport efficiency while designing overall equipment solutions at a lower cost [28].

Displays within the cab allow the operator of the vehicle to view real-time data and information as the vehicle is collecting it. These displays show the current location of the vehicle via GPS, the instantaneous fuel consumption rate, and other performance metrics that help the operator make intelligent decisions in order to maximize the efficiency of the vehicle. The John Deere Gen4® display shows many attributes, such as the instantaneous fuel economy and location of the vehicle within the field, but it also communicates with other vehicles in the same area to share guidance lines, coverage maps, and applied data in order to work the field efficiently [29].

The display associated with Case IH's Advanced Farming System® (AFS®) product, like the Gen4® display, is able to show the location of the vehicle within the field [30]. Using GPS and wireless data networks, it is also possible to check the performance of each vehicle from computers located away from the field. AGCO uses Fuse®, which is much like the Gen4® display and AFS®. Fuse® shows various data on how to improve the efficiency of the specific field operation and includes a seed and dry fertilizer monitoring system, which alerts the operator immediately, via the display, if there is a physical delivery blockage.

Aside from the role CAN-BUS plays in system-to-system communication within a vehicle, the serial network technology has also been integral in the advent of telematics. Telematics is a sector of information technology concerned with how data moves between machines over long distances. Incorporating telematics technology into a vehicle or fleet of vehicles provides the opportunity to utilize collected data outside the scope of an individual machine's operation by integrating it into a server network for wider usage and analysis. While CAN-BUS is not the sole technology responsible for telematics, it serves an important role in communicating large quantities of data that are eventually converted into valuable information for end users [31].

The general architecture of a vehicle telematics system begins with a Telematics Control Unit (TCU), a telematics cloud server, and front-end applications (apps) through which the end user accesses captured data. The TCU is a microcontroller that manages data collection, communication and memory through interfacing with different hardware and software modules. TCUs provide connection ports to CAN-BUS, GPS, General Packet Radio Service, battery, and Bluetooth modules while maintaining a memory unit, a central processing unit, and communication interfaces to wireless fidelity (Wi-Fi), cellular networks, and long-term evolution (LTE) networks [31]. As the central component to a telematics system, the TCU accomplishes the tasks of gathering all the desired data and information from its various connections, synthesizing the information, and communicating to the cloud for use elsewhere. Focusing specifically on the CAN interface, a TCU utilizes the CAN-BUS as a pathway to collect the requested information from the ECUs, as programmed into its operating algorithm. This information acquisition could include any sensor data such as fuel consumption and vehicle speed. By converting the data from the CAN protocols, the TCU can then transfer this data to the telematics cloud server for further post-processing, after which a user would be able to access the data.

The most common usage of telematics across all industries is within fleet management systems. This data collection process allows managers to optimize fuel usage, monitor vehicle downtime, analyze vehicle processes, and track operators driving a specific vehicle [31]. However, different companies also try to bring unique advantages to their telematics packages, which normally materialize in the form of a specialized management software. For construction and forestry equipment, Caterpillar utilizes a company-specific telematics system called ProductLink®, which has both cell and satellite transmission options paired with the user interface VisionLink®. The focuses in these systems include the reduction of idle time and the elimination of catastrophic failures through the reporting of fault codes [32]. John Deere provides customers with the option of a subscription package to the company's telematics network JDLink®, which is customizable to include mobile connections, In-Field Data Sharing®, Operations Center® (where data is synced every 30 seconds to keep it safe and secure), and other features that provide greater connective awareness of interdependent operations

[33]. Case IH takes connectivity to a more automated level with its AFS® product, which has options for autoguidance steering in tractors and combines using AFS AccuGuide® and AFS RowGuide® to aid in year-to-year repeatability. CASE IH's AFS Pro® system monitors several operational metrics and can manage ISOBUS implements [30, 34]. Utilizing CAN-BUS as a communication platform for mobile data transfer has greatly increased the capacity for utilizing data to drive decisions and functions.

In 2009, Agritechnica launched the Isomatch Tellus® VT. This enabled the operator to observe two ISO-BUS machines through one terminal, allowing for the simultaneous control of functions on different platforms. The possible connections to this terminal included a 15-pin ISOBUS, a power connector, an additional 9-pin extension connector, 4 USB interfaces, Bluetooth, internet dongle, EIA-232 port for GPS, and others. Later, software packs such as ISO-XML were added to the VT [11]. Another example of user-focused technology is the Opus A3 CAN-BUS operator panel series from Wachendorff Elektronik, which has two CAN-BUS ports and is specifically designed for outdoor applications that include agricultural machinery [35]. As is evidenced by many of the applications in industry discussed above, different interface technology with CAN-BUS has been important in broadening its usage in a variety of fields. Further discussion of both wireless and non-wireless alternatives to and potential connection points with CAN are explored in the next section.

5.2 ALTERNATIVE CONNECTIVITY AND NETWORKING TO CAN-BUS

Different kinds of interfaces have been specifically developed to allow the conversion of CAN data into a format for Internet of Things (IoT) communication. Two specific technologies of note are CAN-Ethernet and CAN-Bluetooth converters. A CAN-to-Ethernet converter allows the transfer of data in both directions and may be utilized in CAN-BUS monitoring, two-way remote CAN-BUS monitoring, and synchronization [36]. The firmware on such a converter contains both a communication device and a web server. The web server manages the protocol conversions, and the communication device provides the user interface. By combining two CAN-Ethernet converters, two CAN networks can be synchronized, allowing connection between CAN networks on different machines and in remote locations. This may be scaled up further, or a custom software can be programmed to allow the converters to communicate directly to a specific IP address [36].

A CAN-to-Bluetooth gateway, unlike the Ethernet connection, can transfer wireless data directly to a mobile device using classic Bluetooth standards for Android devices and Bluetooth Low Energy for Apple IOS. As with an Ethernet converter, when the devices are used as a pair, a bridge for CAN data can be created for the end user to access [37]. The ISOBlue 2.0 is an example of technology under development that utilizes Bluetooth principles. Currently being researched in the Open Ag Technology and Systems Center (OATS) at Purdue University, ISOBlue 2.0 is an open-source hardware product that connects agricultural machinery to the Cloud [38]. Other interfaces that allow CAN data conversion into different forms have been important tools in making telematics technology viable for off-road agricultural equipment. CAN Logger CLX000, which works between CAN and OBD2, is one such example [39].

Additional wireless technologies that have been used to interface CAN-BUS to IoT devices include Zig-Bee and Wi-Fi. These technologies also function as standalone networks for intra-vehicle and inter-vehicle communication [40]. Similar to the CAN data converters for Bluetooth and Ethernet, ZigBee and Wi-Fi converters have also been utilized to take advantage of their respective benefits in bandwidth, data transfer rate, security, and cost. More detail on each technology's specific advantages is presented in Table 1.1.

TABLE 1.1. *A comparison of wireless technologies capable of interfacing with CAN and IoT devices [41, 43, 44, 45]*

WIRELESS TECHNOLOGY	INSTALLATION COST	BANDWIDTH CAPABILITY	DATA RATE	SECURITY
ZigBee	Medium	Medium	Low	Moderately secure
Bluetooth	Low	Low	Low	Less secure
Wi-Fi	High	High	High	More secure
UWB	Low	High	High	Moderately secure

ZigBee is a globally available wireless networking standard initially created as a home-area network for the control and monitoring of connected devices [41]. ZigBee is beneficial for sensor and vehicle network applications due to its affordable installation and use cost, extensive battery life compared to competing devices, minimal maintenance, security and reliability, and small physical device footprint [41]. ZigBee was built on the IEEE 802.15.4 technical standard, which defines the physical layer and medium access control sublayer for low data-rate wireless personal area networks [42]. CAN-BUS–to–ZigBee conversion has demonstrated benefits in flexibility, convenience, and ease of use in system installation, adding and removing nodes, system updates, and expanded network construction [43].

Wi-Fi is a popular wireless technology for CAN-BUS interfacing and IoT communication. Wi-Fi falls under the IEEE 802.11 standard, which is part of the broader IEEE 802 technical standards for local area networks (LANs) and defines medium access control and physical layer protocols for applying wireless local area network computer communication [46]. This standard also specifies common radio frequency bands that Wi-Fi can communicate on. These include but are not limited to 2.4 GHz, 5 GHz, 6 GHz, and 60 GHz frequencies [46]. Wi-Fi offers a high data rate of up to 54 Mbps and a large bandwidth capability [44]. The most common application for Wi-Fi to CAN-BUS interaction is vehicle-to-cloud telematics services, as discussed in the previous section. On-vehicle Wi-Fi networks also allow for remote control of vehicle systems and provide capability for varying levels of autonomous control. On-vehicle Wi-Fi networks also allow for sending CAN-BUS data from vehicle to vehicle or across several vehicles simultaneously.

Ultra-wideband (UWB) is another wireless technology being researched for vehicle communication systems. UWB is a low-power radio protocol specifically created to improve the location accuracy of wireless technologies. UWB transmits data across a short distance and measures the time it takes for a radio signal to travel between the sending and receiving device [45]. This is similar to the time-of-flight method used with radar. A UWB transmitter sends billions of radio pulses across a wide-spectrum frequency of 7.5 GHz. These pulses are then translated into usable data from a UWB receiver. While UWB is not commonly used in conjunction with CAN-BUS, it has been studied for use in autonomous vehicle navigation and path localization [45].

The continuous development and improvement of autonomous vehicle technology necessitates an increased demand for greater bandwidth and connectivity requirements while still providing an allowance for high system complexity. System complexity in this case could be defined as the added latency from the connected network devices. As many aspects of the interconnected vehicle networks continue to grow, management and network understanding also become more complex. Such aspects include a number of features, routing table configurations, system security, and firewall protections, among other aspects [47]. One of the most promising

TABLE 1.2. *A comparison of CAN characteristics with competing technologies [48]*

NETWORK TYPE	INSTALLATION COST	BANDWIDTH CAPABILITY	SYSTEM COMPLEXITY	FAULT TOLERANCE
CAN	Medium	Low	High	High
FlexRay	High	Medium	High	Medium
MOST	Medium	Medium	Medium	Low
Ethernet	Low	High	Medium	Low

alternatives to vehicle CAN networks are automotive Ethernet-based networks. The market for automotive Ethernet is expected to increase by 22% from 2019 to 2026 [48]. High bandwidth capabilities and improved cost efficiency are two major benefits to automotive Ethernet networks. Instead of a priority-based protocol, Ethernet utilizes a carrier sense multiple access with collision detection strategy [49]. This defines the appropriate device response when multiple control units simultaneously attempt to use a data channel and encounter a data collision. Susceptibility to radio frequency interference, the inability to provide latency at very high frequency, and synchronization issues between timing devices are potential challenges with automotive Ethernet network implementation [48]. Currently, the primary consumption of Ethernet technology in vehicles is enabling personal use of the internet. Ethernet provides rapid data transfer speed, making it ideal for data-intensive applications. However, Ethernet does not adapt well to internal failure, as seen in Table 1.2. A potential associated cost with Ethernet demand increase is the expensive coated wiring needed to provide such high bandwidths.

One type of automotive network communication protocol is FlexRay. FlexRay is a network standard for automotive systems, based on a flexible high data transmission rate high-speed bus system, such as CAN FD [48]. FlexRay is designed for communication of efficiency-type applications in the vehicle. This is due to FlexRay's high complexity allowance and bandwidth. At 10 Mbps on two dual channels, FlexRay can provide up to 20 Mbps of bandwidth, making it optimal for systems such as steering and brakes. CAN shows advantages over FlexRay primarily in cost and error handling [50]. Due to FlexRay's robust complexity and bandwidth, its cost is far greater than CAN on a value per baud rate basis. Although CAN does not generate data transfer rates as fast as FlexRay, it is better suited for the majority of smaller jobs at a far lower cost [50].

Another type of automotive network is Media Oriented System Transport (MOST). MOST provides very fast data transfer at over 24 Mbps. This is because the system was designed to transfer media information within luxury cars, such as GPS, radio, and video systems. MOST has comparable speeds to Ethernet and is more common in automotive applications. However, it handles much less system complexity than CAN and FlexRay, limiting its potential applications [51]. MOST is equipped with plastic optical fiber in its physical layer, which limits electromagnetic interference, thus providing faster speeds and significantly less signal jitter. CAN and MOST have comparable costs, but CAN is better suited for more versatile and sophisticated operations [48].

Overall, CAN shows the most versatility of these four main alternative systems. FlexRay is useful for safety systems due to its high complexity allowance and multiple channel scheme, but it is a higher-cost system by a significant margin. MOST provides one of the best options for media and information transmission, with a faster data transfer rate than two of the other technologies reviewed [50]. However, MOST cannot be used

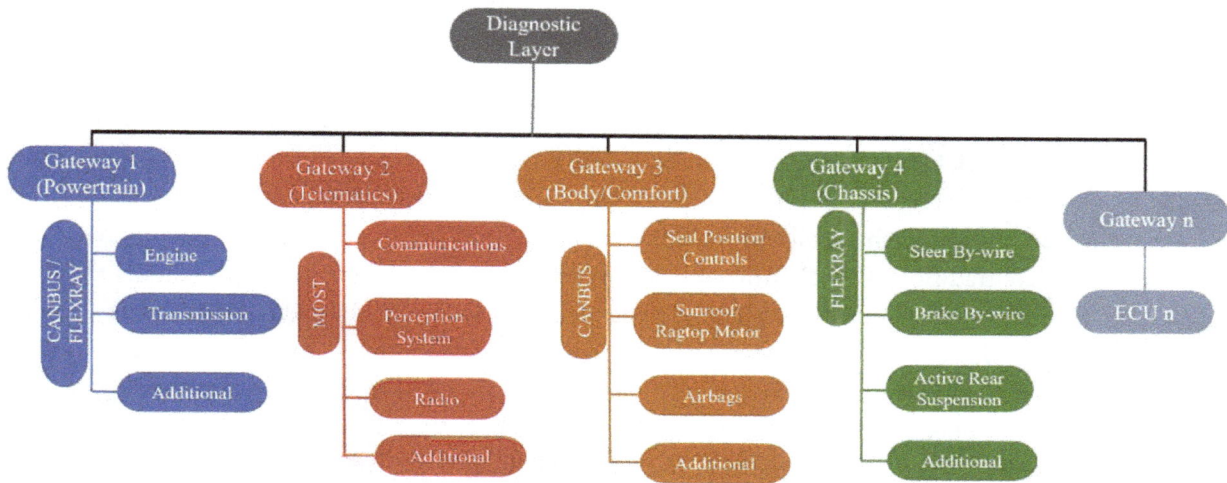

FIGURE 1.6. An example of a FlexRay application.

for highly complex systems. Ethernet provides the fastest data transmission speeds of all the options compared, but it is limited by low complexity allowance and adaptability. CAN, while moderately priced, shows high adaptability to complex systems while providing useful data transfer in a variety of applications [48]. An example of an interconnected system utilizing these networks in a passenger vehicle is shown in Figure 1.6.

5.3 PROSPECTIVE AREAS FOR CAN TECHNOLOGY INCLUSION

Currently, CAN-BUS is used in autonomous vehicle development to gather data from all electronic control sensors and consolidate it onto a single network. By gathering the data into a unified structure, the overall system controller can easily make decisions that affect multiple subsystems at once. This data availability, combined with swift processing, is a key component in the safe operation of autonomous vehicles on the open road and off-road. This centralized system data stream allows for advanced control of smart engine sensors, which provide more efficient management processes. The data handling capability of smart controllers is still an area in need of concentrated improvement. Present research is looking into robust solenoids and other embedded sensors to control valve timing, coolant flow rate, compression ratio, and other key processes in engine operation [52]. Integrated development of these smart controllers with CAN will be crucial to ensuring the safety of autonomous vehicle function execution and travel.

While large-scale agricultural mechanization has been associated with various negative environmental impacts, from soil compaction to harmful exhaust emissions, the advent of digital agriculture has played a key role in increased efficiencies and technological progress within the farming sector, reducing those detrimental elements. The utilization of CANs for improved operation is a research area where further development could have a significant impact with respect to environmental effects. For example, some of the most common technologies for limiting emissions have associated environmental costs that detract from ecological benefit. Although exhaust gas recirculation decreases NO_x emissions, it simultaneously increases specific fuel consumption to lower engine efficiency. Similarly, the post-combustion treatment selective catalytic reduction results in better emissions efficiency but consumes a urea solution that increases freshwater eutrophication risks [53].

Since fuel consumption is primarily dependent on engine speed and torque, it is possible to reliably decrease emissions with the application of alternative driving techniques optimally suited to specific drivetrain design and implement load [54]. However, the plausibility of deriving accurate efficiency metric assessments is limited due to present data scarcity. Current methods for life-cycle assessment studies provide unreliable results because average conditions, such as soil texture, field shape, soil moisture, implement transfer difference, and engine features, have traditionally been utilized in lieu of actual conditions to estimate environmental effects [55]. CAN is advantageously positioned to help address both the data deficiency and inadequate life-cycle assessment techniques due to its data collection and communication strengths. It is possible, for example, that performance metrics could be improved through intelligent sensor solutions that can measure slippage and soil compaction at the wheels of a vehicle and attached implement [13, 54]. These sensors could communicate with sensors in the drivetrain to adjust the effective gearing ratio in real time, reducing soil compaction and preserving the long-term viability of the soil.

An example of an instrument that, when paired with CAN-BUS communication, could be useful in achieving such operational efficiency objectives are inertial measurement units (IMUs). An IMU functions as a sophisticated accelerometer/gyroscope combination. The IMU boasts near zero drift between different operating conditions, and its use of magnetic fields allows it to double as an "electronic compass." The IMU allows for communication across many different CAN-BUS networks to help the tractor and any vehicle make decisions about how to alter the driving style for the terrain to limit "dynamic pitch and roll" through open system communication [52]. While this specific system is not currently implemented on tractors and other off-road vehicles, there is room for its introduction in the emerging field of agricultural autonomy.

Smart agriculture and digital farming practices have gained popularity in the previous decade. These techniques are precursors to a transformative implementation of information technology in the farming world. Going forward, more advanced software systems will use information collected from CAN communication devices to aid in the optimization of machinery designs and more accurate load, use-profile, and duty cycle representations of vehicles and implements [18]. Future applications for CAN-BUS technology include IoT, edge computing, and swarm machinery automation as well as complex control of electrical and electric-hybrid machinery.

IoT implementation in the agricultural sector has gained enormous traction in recent years as a result of its high potential for cross-brand interoperability, scalability, and traceability. The different types of IoT tools being applied are continuing to evolve, increasing the overall adaptability and variety of available systems to end users [56]. IoT systems are currently being implemented on vehicles from John Deere, Case New Holland, AGCO, and others. Future IoT device use on agricultural equipment will likely be in conjunction with multiple onboard network systems. Local storage or cloud computing will be necessary to store and process the vast amount of data created by this potential technology [57]. Data processing aboard the vehicle, near the working equipment, is referred to as "edge computing" (EC) [56, 58]. It is highly probable that agricultural vehicles will eventually be able to perform a variety of complex agronomic tasks from a preprogrammed routing structure through the combined utilization of both IoT and EC technologies.

In addition to on-vehicle IoT technologies, it is probable that field embedded (or in-situ) IoT sensors will also be able to communicate with larger on-farm networks [59]. Several of the previously discussed network configurations are possible whole-farm network options. These include cellular (4G, 5G, and beyond), Wi-Fi,

ZigBee, and UWB. For example, real-time soil moisture can be obtained from field-based connected sensors to create a variable-rate prescription map [60]. Utilized in conjunction with mobile soil penetrometer readings, an accurate map of soil compaction risk can be created. This could allow farmers to tailor their tillage operations to specific areas of the field as well as control vehicle traffic.

Cutting-edge networking research is also being done with robotic and swarm machinery automation [61]. IoT technologies and improved connectivity will allow for the introduction of robotic swarm farming techniques. Swarm farming incorporates multiple small-scale robotic platforms that perform farming operations autonomously in place of larger manned agricultural equipment. This farming strategy, paired with a predetermined path-planning algorithm optimizing how the machines will navigate throughout the field, could allow for near-continuous field operation. Additional benefits could include a centralized command center that is controlled by a single system manager and a significant reduction in the need for skilled labor [62]. The possibility of substituting the modular vehicle design within swarm farming for traditional larger equipment will depend on cost, comparative system productivity, and accuracy. Farmers will demand a significant return on investment and the reliability that they have come to expect from their current machinery. A potential difficulty for CAN-based systems is the large bandwidth requirement for incoming and streaming data. Another potential challenge involves communication protocol differences between traditional CAN-BUS data and more memory-intensive data collected from advanced machine systems, such as perception engines and central processor-based codes [63]. Future developments in CAN-BUS technology should focus on addressing these weaknesses to improve adaptability to upcoming applications.

A major concern in the future of agricultural CAN use, machinery networking, and machine system automation is cybersecurity. Although increased digitization, automation, and precision services have tremendous potential to establish sustainability and profitability in farming systems, the influx of interconnected information technology simultaneously opens the market up to new areas of susceptibility, security risks, and potential targeted cyber attacks [58]. Mission-critical systems are becoming more reliant on internet connectivity, such as controlling farming implements remotely through the ISOBUS with linked management software. LANs have become a requirement in smart farming to enable system/device access to the data and services that control their functions [64]. This increased dependence of agricultural operations on cyber-physical systems has led to the development of novel threats and challenges that can be analyzed in two categories: information technology and agricultural production [58].

From an informational technology standpoint, some of the main threats are unauthorized access of resources/databases under use of falsified identity, interception of node data transfer, facility damage or downtime, malicious data attacks from malware, and compromised control systems to negatively impact decision-making [58]. Due to the nature of modern networked food systems, targeted or accidental disruption of time-sensitive agricultural processes could have a significant economic impact on a global scale. The threat of a concentrated hack on the agricultural sector has become more tangible with the analysis of cybersecurity breaches in recent years, such as the 2017 infrastructure meltdown of Maersk shipping [65]. The vulnerability of wireless local area networks to direct cyber attacks is already a generally recognized problem across all industries [66]. Denial-of-service attacks, as have been demonstrated in the research of Sontowski et al., disrupts in-field sensors and obstruct device network connectivity in smart farm operations [67].

Although the hacking activities of malicious actors is a highlighted concern in cybersecurity, there are also a number of risks associated with agricultural production that stem from physical layer vulnerabilities and limited user knowledge. The harsh environment in which agricultural equipment is used (including extreme weather conditions, dust concentration, and highly variable humidity/temperature fluctuation) can cause power failures or sensor damage [64]. Technology signal interference from other agricultural equipment, such as the high-voltage pulses from solar insecticidal lamps, can also lead to malfunctions and data loss [58].

However, one of the most common threats to cybersecurity is inadequate adoption of safety procedures by farmers who lack full awareness of device functionality. As indicated in research conducted by Nikander et al, farmers are often ill-equipped vis-à-vis time and resources to build LANs with appropriate network equipment, topology expansion planning, and protection software/hardware [64]. This leads to networks that are at risk of system losses due to hardware issues and human error. The adoption of countermeasures to security risks, such as authentication and access control, cryptography, key management, and intrusion detection systems, is dependent on end users understanding the importance of cybersecurity and better fail-safe mechanisms within hardware [58, 64]. These concerns highlight the importance of advancing security protocols in CAN-BUS technology, and it is likely that this will be a targeted focus in the future of CAN developments.

6. CONCLUSION

Key points from this chapter included the following:

- CAN-BUS has played a major role in industry since its debut in 1987 for its groundbreaking use of distributed network principles.
- The establishment of international societies and standards positioned CAN-BUS as the leading serial network system in all vehicles.
- CAN-BUS provides efficient and dependable communication pathways through front and back end context in messaging, error confinement, higher-layer protocols, and subsystem differentiation.
- CAN-BUS has revolutionized data collection and analysis in multiple industries, especially in the agricultural sector.
- When paired with wired and wireless technologies, CAN is an advantageous communication pathway for expanding the reach of data communication beyond point source limitations.
- Challenges for future CAN iterations include increasing bandwidth and security measures while decreasing latency and hardware vulnerabilities.

This chapter has reviewed CAN-BUS technology, including its invention, early applications, fundamentals, and standards development. Early applications of CAN-BUS came from European car manufacturers, which incorporated electronic control units for engine care management. The development of standards to allow consistent communication methods in CAN-BUS technology, such as ISO 11898, SAE J1939, and ISO 11783, were important for allowing serial networks to be applied in multiple vehicle types and industries. Modern-day uses, alternative connectivity and networks, and potential future applications have also been examined. CAN technology is responsible for the transmission, logging, and analysis of engine and machine system data currently used by vehicle manufacturers. Understanding CAN-BUS communication protocols provides insight into the advantages, uses, and future evolutions of distributed control networks.

CAN-BUS technology fundamentals, such as physical and data message structures, components, error handling, and message channel support, are useful in understanding the strengths and limitations of CAN technology. Through the use of high- and low-speed CAN-BUS configurations, arbitration codes, and broadcast-style communication, CAN-BUS can efficiently and reliably transfer messages across a vehicle's control system to ensure accurate, real-time data communication. As electronic connectivity has increased the sophistication of off-road vehicle operation management, new applications using CAN with external networks have been an important area of communications advancement within the agricultural sector. The development of converters between CAN data and other wireless data types has been important in keeping CAN-BUS integrated and relevant in the vehicle fleet telematics expansion. More research into wireless CAN may be an important direction for serial network technology going forward.

Specific CAN-BUS applications in ongoing autonomous vehicle development research include component data consolidation, embedded sensors, IoT devices, and machine-to-machine communication strategies. Future technologies that might benefit CAN-BUS technology by their incorporation include local-to-cloud data transmission, autonomous swarm vehicle management, and increased cybersecurity protocols. Although CANs face limitations within both bandwidth and latency, they still function as effective inputs to more advanced vehicle systems and more sophisticated remote networks. The potential of CAN-BUS technologies has clearly not been fully exhausted, and these technologies will continue to play an important role in the advancement of agricultural machinery and farming practices.

7. ACKNOWLEDGMENTS

We would like to acknowledge our fellow classmates from Dr. Stwalley's fall 2020 Off-Highway Vehicle Design class at Purdue University's School of Agricultural and Biological Engineering for their contributions to the structure and content of this technical chapter.

8. CONFLICT OF INTEREST

The authors declare no conflict of interest.

9. CAN-BUS QUESTIONS

1. Describe how the generalized CAN-BUS architecture was designed so that it can transmit data with shorted or broken wires.
2. If a CAN-BUS reaches its maximum message-carrying capacity at roughly 35% of its theoretical capacity, how is it possible to continue to expand the network beyond that point?
3. Describe the advantages of a CAN-BUS network over Ethernet for mobile equipment.
4. List five advantages attributable to CAN-BUS compared to traditional point-to-point electronic harnesses. List three overall system disadvantages.
5. Describe how collected, processed, or acted upon CAN-BUS data can be recorded or uplinked beyond the mobile machine network.

10. REFERENCES

[1] Copperhill Technologies, "A Brief Introduction to Controller Area Network," 2020a. [Internet]. Available: https://copperhilltech.com/a-brief-introduction-to-controller-area-network/, [Last accessed: May 15, 2021].

[2] CiA, "History of CAN Technology," 2018. [Internet]. Available: https://www.cancia.org/canknowledge/can/can-history/, [Last accessed: May 15, 2021].

[3] X. Gao, D. Huang, Y. Chen, W. Jin, and Y. Luo, "The Design of a Distributed Control System Based on CAN Bus," IEEE International Conference on Mechatronics and Automation, pp. 1118–1122, 2013, https://doi.org/10.1109/ICMA.2013.6618071.

[4] W. Voss, *A Comprehensible Guide to Controller Area Network.* Copperhill Technologies Corporation, 2011.

[5] CSS Electronics, "CAN Bus Explained—a Simple Intro," 2020a. [Internet]. Available: https://www.csselectronics.com/screen/page/simple-intro-to-can-bus/language/en, [Last accessed: May 15, 2021].

[6] Kvaser, "The Kvaser Dictionary for Help in Understanding CAN Bus Systems," 2020. [Internet]. Available: https://www.kvaser.com/about-can/can-dictionary/, [Last accessed: May 15, 2021].

[7] S. Maradana, "CAN Basics," 2012. [Internet]. Available: https://automotivetechis.wordpress.com/2012/06/01/can-basics-faq/, [Last accessed: May 15, 2021].

[8] ISO, "About Us," 2020. [Internet]. Available: https://www.iso.org/about-us.html, [Last accessed: May 15, 2021].

[9] CiA, "CAN-Based Higher-Layer Protocols (HLP)," 2017. [Internet]. Available: https://www.can-cia.org/can-knowledge/hlp/higher-layer-protocols/, [Last accessed: May 15, 2021].

[10] ISO, "ISO 11898–1:2015," 2015. [Internet]. Available: https://www.iso.org/standard/63648.html, [Last accessed: May 15, 2021].

[11] CiA, "Isobus—The CAN-Based Network System," CAN Newsletter, pp. 44–48, 2010.

[12] C. Goering, M. Stone, D. Smith, and P. Turnquist, "Off-Road Vehicle Engineering Principles," American Society of Agricultural Engineers, 2006, ISBN: 978–1892769268.

[13] F. S. Al-Aani, "CAN Bus Technology for Agricultural Machine Management," Iowa State University Capstones, Theses, and Dissertations, 2019.

[14] M. L. Stone, K. D. McKee, C. W. Formwalt, and R. K. Benneweis, "ISO 11783: An Electronic Communications Protocol for Agricultural Equipment." Agricultural Equipment Technology Conference, pp. 1–17, February 1999.

[15] Kvaser, "CAN Bus Error Handling," 2017. [Internet]. Available: https://www.kvaser.com/about-can/the-can-protocol/can-error-handling/, [Last accessed: May 15, 2021].

[16] CSS Electronics , "CANopen Explained—A Simple Intro," 2020. [Internet]. Available: https://www.csselectronics.com/screen/page/canopen-tutorial-simple-intro/language/en, [Last accessed: May 15, 2021].

[17] Kvaser, "Higher Layer Protocols," [Internet]. Available: https://www.kvaser.com/about-can/higher-layer-protocols/, [Last accessed: May 15, 2021].

[18] R. Rohrer, S. Pitla, and J. Luck, "Tractor CAN Bus Interface Tools and Application Development for Real-time Data Analysis," *Computers and Electronics in Agriculture,* p. 163, 2019, https://doi.org/10.1016/j.compag.2019.06.002.

[19] A. Polcar, J. Čupera, and V. Kumbár, "Calibration and Its Use in Measuring Fuel Consumption with the CAN-Bus Network," *Acta Universitatis Agriculturaevet Silviculturae Mendelianae Brunensis,* vol. 64, no. 2, pp. 503–507, 2016, https://doi.org/10.11118/actaun201664020503.

[20] ISO, "ISO 11898-2:2016," 2016. [Internet]. Available: https://www.iso.org/standard/67244.html, [Last accessed: May 15, 2021].

[21] ISO, "ISO 11898-3:2006," 2006. [Internet]. Available: https://www.iso.org/standard/36055.html, [Last accessed: May 15, 2021].

[22] CSS Electronics, "J1939 Explained—A Simple Intro," 2020b. [Internet]. Available: https://www.csselectronics.com/screen/page/simple-intro-j1939-explained, [Last accessed: May 15, 2021].

[23] W. Voss, "Guide to SAE J1939—J1939 Message Format," 2018. [Internet]. Available: https://copperhilltech.com/blog/guide-to-sae-j1939-j1939-message-format/, [Last accessed: May 15, 2021].

[24] Copperhill Technologies, "A Brief Introduction to the SAE J1939 Protocol," 2020b. [Internet]. Available: https://copperhilltech.com/a-brief-introduction-to-the-sae-j1939-protocol/, [Last accessed: May 15, 2021].

[25] W. Voss, "SAE J1939 Bandwidth, Busload and Message Frame Frequency," 2019. [Internet]. Available: https://copperhilltech.com/blog/sae-j1939-bandwidth-busload-and-message-frame-frequency/, [Last accessed: May 15, 2021].

[26] A. Brehmer, "CAN Based Protocols in Avionics," 2014. [Internet]. Available: https://ieeexplore.ieee.org/stamp/stamp.jsp?arnumber=6979561, [Last accessed: May 15, 2021].

[27] D. Fanton, "Why All the Fuss about CAN Bus?," 2020. [Internet]. Available: https://www.onlogic.com/company/io-hub/fuss-can-bus/, [Last accessed: May 15, 2021].

[28] M. J. Darr, "CAN Bus Technology Enables Advanced Machinery Management," Agricultural and Biosystems Engineering Publications, p. 329, 2012.

[29] John Deere, "Gen 4 Command Center Display," 2020. [Internet]. Available: http://www.deere.com/en/technology-products/precision-ag-technology/guidance/gen-4-premium-activation/, [Last accessed: May 15, 2021].

[30] CASE IH, "AFS," 2020a. [Internet]. Available: https://www.caseih.com/apac/en-int/products/advanced-farming-system, [Last accessed: May 15, 2021].

[31] Embitel, "Technology behind Telematics Explained: How Does a Vehicle Telematics Solution Work?," 2018. [Internet]. Available: embitel.com/blog/embedded-blog/tech-behind-telematics-explained-how-does-a-vehicle-telematics-solution-work, [Last accessed: May 15, 2021].

[32] Caterpillar, "Telematics—Your Link to Equipment Cost Savings," 2020. [Internet]. Available: https://www.cat.com/en_US/articles/customer-stories/forestry/telematics.html, [Last accessed: May 15, 2021].

[33] John Deere, "Operations Center," 2020. [Internet]. Available: https://www.deere.com/en/technology-products/precision-ag-technology/data-management/operations-center, [Last accessed: May 15, 2021].

[34] CASE IH, "AFS Connect," 2020. [Internet]. Available: https://www.caseih.com/apac/en-int/products/advanced-farming-system/afs-connect, [Last accessed: May 15, 2021].

[35] Topcon, "Technical Data Sheet OPUS A3 ECO Full," 2019. [Internet]. Available: http://www.motronica.com/wp-content/uploads/2019/08/TDS_A3e_FULL.pdf, [Last accessed: May 15, 2021].

[36] Axiomatic Technologies Corporation, "Technical Datasheet #TDAX140900," 2020. [Internet]. Available: https://www.axiomatic.com/TDAX140900.pdf, [Last accessed: May 15, 2021].

[37] Axiomatic Technologies Corporation, "Technical Datasheet #TDAX141100," 2018. [Internet]. Available: https://www.axiomatic.com/TDAX141100.pdf, [Last accessed: May 15, 2021].

[38] B. Wallheimer, "Purdue Partnering on 5G Research to Improve Ag Automation," 2019. [Internet]. Available: https://ag.purdue.edu/stories/purdue-partnering-on-5g-research-to-improve-ag-automation/, [Last accessed: May 15, 2021].

[39] CSS Electronics, "OBD2 Explained—a Simple Intro," 2020c. [Internet]. Available: https://www.csselectronics.com/screen/page/simple-intro-obd2-explained, [Last accessed: May 15, 2021].

[40] S. Dorle, D. Deshpande, A. Keskar, and M. Chakole, "Vehicle Classification and Communication Using Zigbee Protocol," 3rd International Conference on Emerging Trends in Engineering and Technology, 2010, pp. 106–109, https://doi.org/10.1109/CETET.2010.170.

[41] Silicon Labs, Inc., "ZigBee-Based Home Area Networks Enable Smarter Energy Management," n.d., pp. 1–4. https://www.silabs.com/documents/public/white-papers/ZigBee-based-HANs-for-Energy-Management.pdf.

[42] Institute of Electrical and Electronics Engineers, "IEEE 802.15.4-2020—IEEE Standard for Low-Rate Wireless Networks," 2020. https://doi.org/10.1109/IEEESTD.2020.9144691.

[43] Y. Li, C. Yu, and H. Li, "The Design of ZigBee Communication Convertor Based on CAN," International Conference on Computer Application and System Modeling (ICCASM 2010), 2010, ISBN: 978-1-4244-7237-6/10.

[44] N. Chhabra, "Comparative Analysis of Different Wireless Technologies," International Journal of Scientific Research in Network Security and Communication, vol. 1, no. 5, pp. 13–17, 2013, 2321–3256.

[45] J.-S. Lee, Y.-W. Su, and C.-C. Shen, "A Comparative Study of Wireless Protocols: Bluetooth, UWB, ZigBee, and Wi-Fi," IECON 2007: 33rd Annual Conference of the IEEE Industrial Electronics Society, 2007 https://doi.org/10.1109/IECON.2007.4460126.

[46] Institute of Electrical and Electronics Engineers, "IEEE 802.11–2016—IEEE Standard for Information Technology—Telecommunications and Information Exchange between Systems Local and Metropolitan Area Networks—Specific Requirements—Part 11," 2016.

[47] M. Behringer and M. Kuehne, "Network Complexity and How to Deal with It," 2011. [Internet]. Available: https://labs.ripe.net/Members/mbehring/network-complexity-and-how-to-deal-with-it, [Last accessed: December 20, 2020].

[48] Navixy, "CAN Bus and Alternatives," 2020. [Internet]. Available: https://www.navixy.com/docs/academy/can-bus/can-and-alternatives/, [Last accessed: May 15, 2021].

[49] A. Quine, "Carrier Sense Multiple Access Collision Detect (CSMA/CD) Explained," IT Professional's Resource Center (ITPRC), 2008. [Internet]. Available: https://www.itprc.com/carrier-sense-multiple-access-collision-detect-csmacd-explained/. [Last accessed: May 8], 2021.

[50] National Instruments Corp., "FlexRay Automotive Communication Bus Overview," National Instruments Corp., 2020. [Internet]. Available: https://www.ni.com/en-us/innovations/white-papers/06/flexray-automotive-communication-bus-overview.html. [Last accessed: May 8, 2021].

[51] S. Tuohy, M. Glavin , C. Hughes, E. Jones, M. Trivedi, and L. Kilmartin, "Intra-Vehicle Networks: A Review," IEEE Transactions of Intelligent Transportation Systems, pp. 1–12, 2014, https://doi.org/10.1109/TITS.2014.2320605.

[52] M. Horton, "What Can a CANbus IMU Do to Make an Autonomous Vehicle Safer?," 2019. [Internet]. Available: https://www.autonomousvehicleinternational.com/opinion/what-can-a-canbus-imu-do-to-make-an-autonomous-vehicle-safer.html, [Last accessed: May 15, 2021].

[53] J. Bacenetti, D. Lovarelli, D. Facchinetti, and D. Pessina, "An Environmental Comparison of Techniques to Reduce Pollutants Emissions Related to Agricultural Tractors," Biosystems Engineering, vol. 171, pp. 30–40, 2018, https://doi.org/10.1016/j.biosystemseng.2018.04.014.

[54] M. Lindgren and P.-A. Hansson, "PM-Power and Machinery: Effects of Engine Control Strategies and Transmission Characteristics on the Exhaust Gas Emissions from an Agricultural Tractor," Biosystems Engineering, vol. 83, no. 1, pp. 55–65, 2002, https://doi.org/10.1006/bioe.2002.0099.

[55] D. Lovarelli and J. Bacenetti, "Bridging the Gap between Reliable Data Collection and Environmental Impact for Mechanised Field operations," Biosystems Engineering, vol. 160, pp. 109–123, 2017, https://doi.org/10.1016/j.biosystemseng.2017.06.002.

[56] A. Kamilaris, F. Gao, F. Prenafeta-Boldum and M. Ali, "Agri-IoT: A Semantic Framework for Internet of Things–Enabled Smart Farming Applications," 2016 IEEE 3rd World Forum on Internet of Things, pp. 442–447, 2016, https://doi.org/10.1109/WF-IoT.2016.7845467.

[57] R. Morabito, V. Cozzolino, A. Y. Ding, N. Beijar, and J. Ott, "Consolidate IoT Edge Computing with Lightweight Virtualization," IEEE Network, pp. 102–111, 2018, https://doi.org/10.1109/MNET.2018.1700175.

[58] X. Yang, L. Shu, J. Chen, M. A. Ferrag, J. Wu, E. Nurellari, and K. Huang, "A Survey on Smart Agriculture: Development Modes, Technologies, and Security and Privacy Challenges," IEEE/CAA Journal of Automatica Sinica, vol. 8, no. 2, pp. 273–302, 2021, https://doi.org/10.1109/JAS.2020.1003536.

[59] J. Liang, X. Liu, and K. Liao, "Soil Moisture Retrieval Using UWB Echoes via Fuzzy Logic and Machine Learning," IEEE Internet of Things Journal, vol. 5, no. 5, pp. 3344–3352, 2018, https://doi.org/10.1109/JIOT.2017.2760338.

[60] A. P. Atmaja, A. E. Hakim, A. P. A. Wibowo, and L. A. Pratama, "Communication Systems of Smart Agriculture," Journal of Robotics and Control, vol. 2, no. 4, pp. 297–301, 2021, https://doi.org/10.18196/jrc.2495.

[61] D. Albiero, A. P. Garcia, C. K. Umezu, and R. L. de Paulo, "Swarm Robots in Agriculture," CoRR, vol. abs/2103.06732, 2021, https://doi.org/10.48011/asba.v2i1.1144.

[62] A. Sharda, "Is Swarm Farming the Future of Farming?," 2020. [Internet]. Available: https://www.precisionagreviews.com/post/is-swarm-farming-the-future-of-farming, [Last accessed: May 15, 2021].

[63] J. Čupera and P. Sedlák, "The Use of CAN-Bus Messages of an Agricultural Tractor for Monitoring Its Operation," Research in Agricultural Engineering, vol. 57, no. 4, pp. 117–127, 2011, https://doi.org/10.17221/20/2011-RAE.

[64] J. Nikander, O. Manninen, and M. Laajalahti, "Requirements for Cybersecurity in Agricultural Communication Networks," Computers and Electronics in Agriculture, vol. 179, pp. 1–10, 2020, https://doi.org/10.1016/j.compag.2020.105776.

[65] Jahn Research Group, "Cyber Risk and Security Implications in Smart Agriculture and Food Systems," Madison, 2019. https://jahnresearchgroup.webhosting.cals.wisc.edu/wp-content/uploads/sites/223/2019/01/Agricultural-Cyber-Risk-and-Security.pdf.

[66] D. Kraus, E. Leitgeb, T. Plank, and M. Löschnigg, "Replacement of the Controller Area Network (CAN) Protocol for Future Automotive Bus System Solutions by Substitution via Optical Networks," *18th International Conference on Transparent Optical Networks (ICTON)*, pp. 1–8, 2016, https://doi.org/10.1109/ICTON.2016.7550335.

[67] S. Sontowski, M. Gupta, S. S. L. Chukkapalli, M. Abdelsalam, S. Mittal, A. Joshi, and R. Sandhu, "Cyber Attacks on Smart Farming Infrastructure," UMBC, Baltimore, 2020, https://doi.org/10.1109/CIC50333.2020.00025.

CHAPTER 2. SOIL COMPACTION

Soil compaction represents one of the most intractable problems in modern agriculture. It is impossible to move animals or equipment on the surface of a field without creating soil compaction, yet crops cannot be grown on any farm without the use of some form of technology moving over the soil. Sustainable solutions for soil compaction issues require an understanding of the phenomena and the techniques to minimize the problems associated with the use of soil media. Understanding the difference between tire and track performance is vital to designing new off-road equipment. Recognizing the effects of varying types of tillage equipment on different soils as well as crop rotations and overall farming protocols is essential to maintaining the health of a farm's soils over time. Agricultural engineers must understand the nuances of operating on a compressible media as well as how to mitigate past problems that might be diminishing the ground's potential. The overview provided here was originally published in *Sustainable Crop Production: Recent Advances* and presents the compressible media issues associated with modern agricultural production techniques as well as the current best practices to lessen the harmful effects of compaction on the sustainability of the soil.

REDUCING SOIL COMPACTION FROM EQUIPMENT TO ENHANCE AGRICULTURAL SUSTAINABILITY

MICHAEL M. BOLAND, PURDUE UNIVERSITY AGRICULTURAL & BIOLOGICAL ENGINEERING

YOUNG U. CHOI, PURDUE UNIVERSITY AGRICULTURAL & BIOLOGICAL ENGINEERING

DANIEL G. FOLEY, PURDUE UNIVERSITY AGRICULTURAL & BIOLOGICAL ENGINEERING

MATTHEW S. GOBEL, PURDUE UNIVERSITY AGRICULTURAL & BIOLOGICAL ENGINEERING

NATHAN C. SPRAGUE, PURDUE UNIVERSITY AGRICULTURAL & BIOLOGICAL ENGINEERING

SANTIAGO GUEVARA-OCANA, PURDUE UNIVERSITY POLYTECHNIC INSTITUTE

YURY A. KULESHOV, PURDUE UNIVERSITY POLYTECHNIC INSTITUTE

ROBERT M. STWALLEY III, PURDUE UNIVERSITY AGRICULTURAL & BIOLOGICAL ENGINEERING

ABSTRACT

The compaction of agricultural soils is a problem that cannot be solved, only managed. As a compressible media, soil is impossible to travel across or till without causing some collapse of the existing structure. A large agricultural vehicle weighting 10 *tons* can create significant compaction over one m in depth when under a draft load. If left uncorrected, farmers can see up to a 50% reduction in yield from long-term compaction. This chapter will describe the effects of soil compaction on the environment, crop quality, and economic sustainability. The base causes of detrimental compaction will be examined, along with the engineering designs for vehicles that minimize the problem. The tracks-versus-tires debate will be thoroughly discussed, and the advantages and disadvantages of each type of undercarriage system will be detailed. It will be shown that although tires represent the likely current best economic option for vehicle support, the potential of tracks to reduce compaction has yet to be fully exploited. The advantages of four-wheel drive vehicles in reducing soil compaction will be shown, along with the mitigation potential of independently driven wheels and active soil interaction feedback loops. The design of crop production tillage equipment and tillage tool working points will be explored, along with the concept of critical tillage depth. Equipment for compaction relief will also be discussed, as will the sustainable agricultural protocols of cover crops, crop rotation, and controlled traffic farming.

KEYWORDS: agricultural tillage, compaction remediation, cover crops, off-road vehicle design, tires, tracks, tractors, soil compaction, sustainability

1. INTRODUCTION

Since the late 1960s, the agricultural industry has taken an increasing interest in the effects of soil compaction on soil health, agricultural practices, water runoff, and the sustainability of grain production. Compaction results from any practice that includes traveling over the soil. This can be caused by heavy-axle machinery, excessive ground working, livestock, or specific geotechnical practices, such as rolling, which is used to compact the soil in preparation for construction. Repeated soil compaction experiences have cumulative negative effects for agricultural soils, such as a decrease in pore space, reduced pore nutrient and water uptake, denitrification, and enhanced difficulties in seed germination. The effects of compaction also extend beyond agriculture and are of concern to environmental specialists all over the world. For instance, high compaction rates increase the likelihood of water retention issues, water runoff, and erosion. From the last 60 years of research, modern agricultural operations have progressed to incorporate a variety of approaches for reducing soil compaction. These approaches include equipment solutions such as tracked implements, happy-seeders, and complex multicrop planters that reduce field traffic. Within the scope of production agriculture, many existing practices unrelated to vehicle design, such as no-till seeding, have decreased the impact of soil compaction and helped to repair damaged and heavily compacted soils. These design improvements and management practices will be explored in this chapter, and their effectiveness will be measured. This topic is particularly timely and relevant because present-day tractors have increased in size compared to traditional row crop tractors for better productivity and field efficiency. Although most smaller-scale agricultural equipment is used for multiple tasks, the presence of a variety of different off-road vehicles on the market indicates a broad need for various equipment types and provides an opportunity for exploring the existing and potential solutions to soil compaction problems in different off-road vehicle designs. The chapter will proceed with an analysis of how soil compaction is addressed in machine design as well as new areas that deserve more specific research and improvement. Multiple factors are involved in soil compaction, and multiple designs exist to address these various factors. The present off-road vehicle offerings are clearly less than ideal for long-term soil health. There is a potential for improvement in existing designs to benefit all involved stakeholders, and this potential will be explored.

2. THE RELEVANCE OF SOIL COMPACTION AND ITS EFFECTS ON SUSTAINABILITY

Interest in soil compaction dates back to the time when humans started to use draft animals as a main source of power in agriculture. Many authors have addressed the problem since the 19th century. One of the first recognitions of the problem in academic literature dates back to 1857, with a description of the Fowler steam engine–powered plowing system [1]. Draft animals are still being used on vast areas of land in developing countries, and the animal-induced compaction problem continues to this day. The growing use of steam-powered tractors added to soil compaction concerns in the second half of the 19th century. While the mass-power ratio allowed for the widespread use of powerful tractors, the vehicles were still very heavy, and the need to minimize wheel impact on the soil was quickly recognized.

Different engineers have attempted to address the problem in multiple ways. These attempts didn't lead to a unified design, but they did move the engineering thinking forward and were instrumental in creating the more successful designs of the 20th century. Between the last decade of the 19th century and 1904, internal combustion engines replaced steam engines on tractors in America, and a new era of agriculture began. The better mass-power ratio of the internal combustion engines provided for lighter designs and less impact on the soil, but other problems emerged. Mass agriculture meant the more intensive use of fields and more impact on the soil per a given period in time. One of the first experiments describing soil compaction problems was run in 1944. The topic has continued to be a strong focus for agricultural researchers. Raney, Edminster, and Allaway conducted the first review of literature on compaction in agricultural soils in America, which included 43 references [2]. The so-called load index started to grow at about the same time and exceeded that of the early 20th century by the 1970s. The soil compaction problem continues to be a major focus of agricultural, industrial, and academic practitioners and researchers. One industrial example is Caterpillar's efforts to use tracks in agriculture to decrease soil compaction [3]. Other modern solutions have attempted to address the problem, and the current review will introduce them to the reader [4–6].

2.1 ENVIRONMENTAL IMPACT OF SOIL COMPACTION

Soil compaction has a measurable influence on the environment, specifically on atmospheric, water, and soil resources. Agricultural operations have a major impact on the atmosphere through the emission of greenhouse gases. Soil "compaction may change the fluxes of these gases from the soil to the atmosphere because of its influence on soil permeability, soil aeration and crop development" [4, p. 8]. Water resources include both surface water and groundwater volumes. Soil compaction negatively affects the infiltration of different substances into the ground. Ammonia injected into the soil can escape into the atmosphere faster in a compacted soil than in an uncompacted one. Soil compaction also perpetuates the accumulation of rainwater on the surface in low parts of the field and increases the likelihood of runoff events. The latter leads to excessive sediment and chemical transfer into surface groundwater resources, such as local rivers, lakes, ponds, and bigger regional natural water reservoirs [5, 7].

Soil biota is responsible for the decomposition of organic matter, the release of nutrients, and the formation of aggregates [4]. Such tasks are performed by microfauna (bacteria, fungi), which are fed upon by meso- and macrofauna (protozoa, nematodes, arthropods) within the soil food web. Figure 2.1 illustrates some of these various interconnections between the living things in the soil.

Soil compaction creates a rearrangement of soil particles that leads to a reduction of void space, a phenomenon that can be measured in several different ways. At first glance, there are visual and tactile methods that can provide a quick assessment, but to quantify the effects of soil compaction, physical parameters must be measured. Direct and indirect measures are used together to enable a deeper understanding of the characteristics of the total volume of the soil, such as bulk density (direct), soil strength (indirect), soil electrical resistivity (indirect), and water infiltration rate (indirect). Figure 2.2 shows two examples of soil profiles exhibiting compaction effects. The soil on the left has a better structure above and below the compacted layer located between a depth of 10 *cm* and 40 *cm* [9]. On the right, a compacted layer in wetland creates a toxic environment for roots and soil biota. Soil compaction effects vary by location based on multiple interconnected factors, making a comprehensive assessment of specific fields the key to securing the sustainability of any agricultural operation over time.

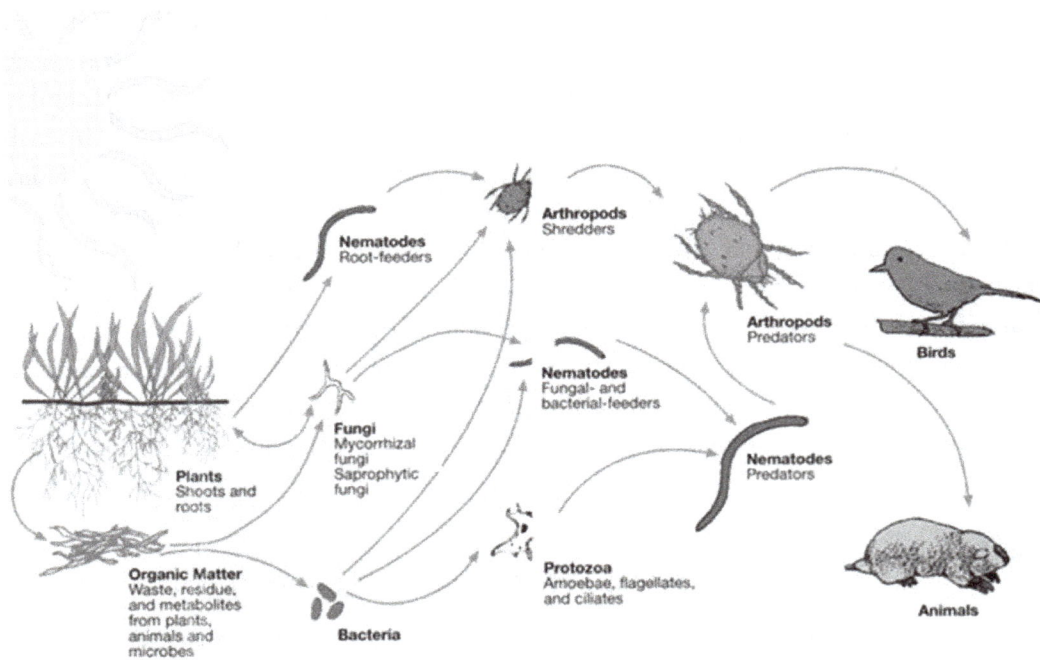

FIGURE 2.1. Soil biota species and food web [8].

FIGURE 2.2. A compacted layer under dryland canola (left), and a grey anaerobic layer in a clay loam soil (right) [9].

2.2 EFFECTS ON HARVEST QUALITY AND FARMLANDS

Soil compaction has a directly visible effect on the crop that is being grown in the degraded area. As soil compacts, it reaches a point of root growth restriction that is highly detrimental to both the quality and health of plants as well as the quantity of the cultivated crop yield [10]. The lack of loose soil aggregates prevents strong root formations. This leaves crops more susceptible to wind and water damage. There is reduced nutrient uptake, since the root mass of the plant is diminished in both absorption volume and effectiveness. Individual plants are less healthy and produce significantly less grain and forage mass. Perennial crops, such as many fruit plants, stop root growth when confronted with significant compaction. Beyond the lack of void space, nutrient uptake in compacted topsoil is greatly reduced as the biological health of the soil diminishes [11]. Crops growing in densified soils can be expected to be brittle due to the reduced nutrient intake, as soil compaction causes reduced aerobic microbial activity and denitrification [12]. Soil compaction has a progressively negative effect on the biosphere. As the soil is compacted and continuously depleted, natural vegetation, such as weeds and grasses, quickly gets restricted from lack of soil aeration. The crushing of the soil and the diminishing amount of additional biomass that would otherwise be introduced into the soil can eventually lead to an elimination of plant life, causing an open and exposed soil surface. Soil in this condition is more easily impacted by wind and water erosion. If preventative measures are not taken, the effects of soil compaction on crop quality and farmland are cumulative and can take place quickly and have lasting damage [13].

As shown by a review of the soil compaction literature [14], studies detailing the continuing long-term effect of compaction on a specific piece of ground are rare. However, as shown in Figure 2.3, it can take years for soil to naturally recover following a single compaction event [15]. Studies in cotton, as displayed in Figure 2.4, show a significant decline in crop yield within the initial season of the compaction event [16]. Since the effects of compaction are cumulative and continue from one season into the next, it can be inferred that the decline from unmitigated soil compaction will continue to grow and magnify under the same management processes. Figure 2.5 presents the general effect over time on production costs and gross margin of the farming operation [14].

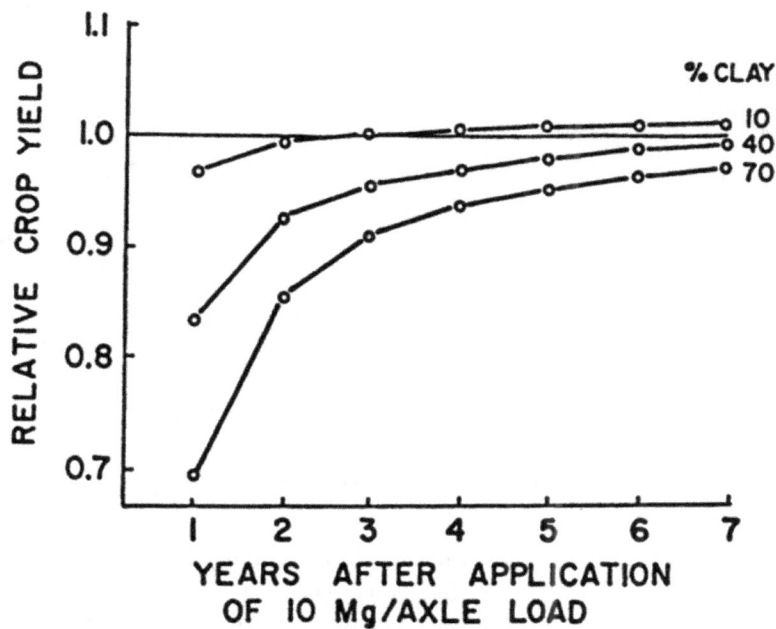

FIGURE 2.3. Yield recovery following a significant compaction event [15].

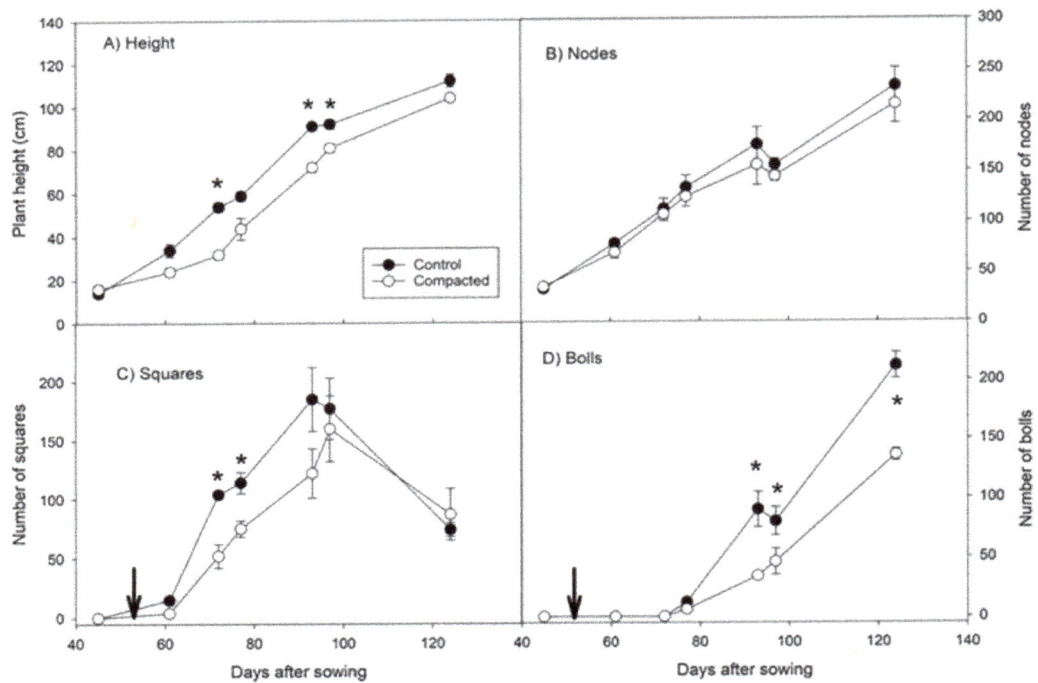

FIGURE 2.4. Difference in same year cotton yield between compacted and uncompacted ground [16].

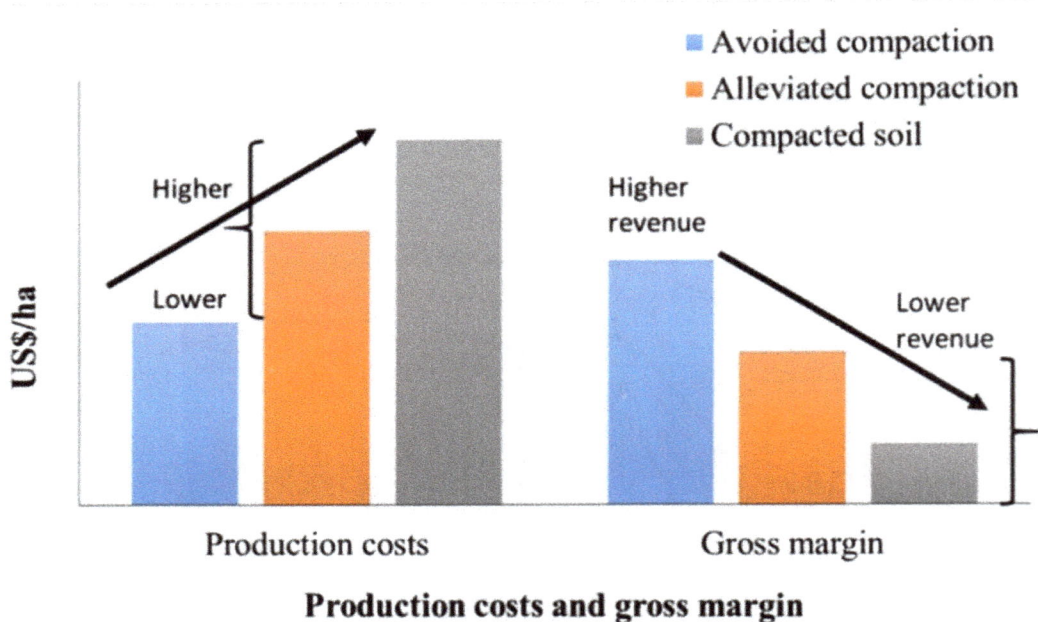

Production costs and gross margin

FIGURE 2.5. The generalized trends for production cost and gross margin for avoided compaction, relieved compaction, and compacted soils in production [14]. In terms of production costs, avoided compaction is the lowest, alleviated compaction is slightly higher, and compacted soil is the highest. In terms of gross margin, the highest revenue correlates with avoided compaction, alleviated compaction is slightly lower, and compacted soil is the lowest.

2.3 SOCIAL AND ECONOMIC IMPACT

Soil compaction has a negative impact on the economy of agricultural operations in the long-term. Soil compaction decreases the quantity and quality of harvest. Continuous and unaddressed soil compaction does not allow soil to sustainably recover through natural means [17, 18]. This affects local food security in the regions where the traditional economy relies on agriculture. The local quality of life and the general economic health of an agricultural region is adversely affected when local soils become compacted. Multiple potential solutions can help minimize the impact. Some solutions come from farmer experience and depend on the operator in the field. Others are industry-wide general practices. Academic researchers model the problem by studying economic impact. These models provide for better forecasting, equipment selection, and targeted problem-solving. One model suggests that in the short-term, the negative impacts of soil compaction can be compensated by "more timely field operations." Profits and productivity generate energy costs, air pollution, capital costs, timeliness costs, and soil erosion, which are also evaluated [19]. Other studies address the problem through even more sophisticated modeling. Additional effort is needed to standardize corporate and academic researchers' efforts to improve these models for specific predictive tasks. Soil compaction models have the potential to help businesses and governments develop advanced solutions for real-world agricultural problems through an improved understanding of social and economic impacts.

The extent of the economic effect of soil compaction is difficult to quantify, as it is ultimately very situational. Under circumstances where soil requires additional operation to alleviate the effects of long-term compaction, the cost of crop production rapidly increases to unviability. Soil health does not always deteriorate to the point where intervention is required, but this does not mean that these farming operations are unaffected. The most common issue caused by soil compaction is the decrease in crop productivity. Figure 2.6 summarizes the

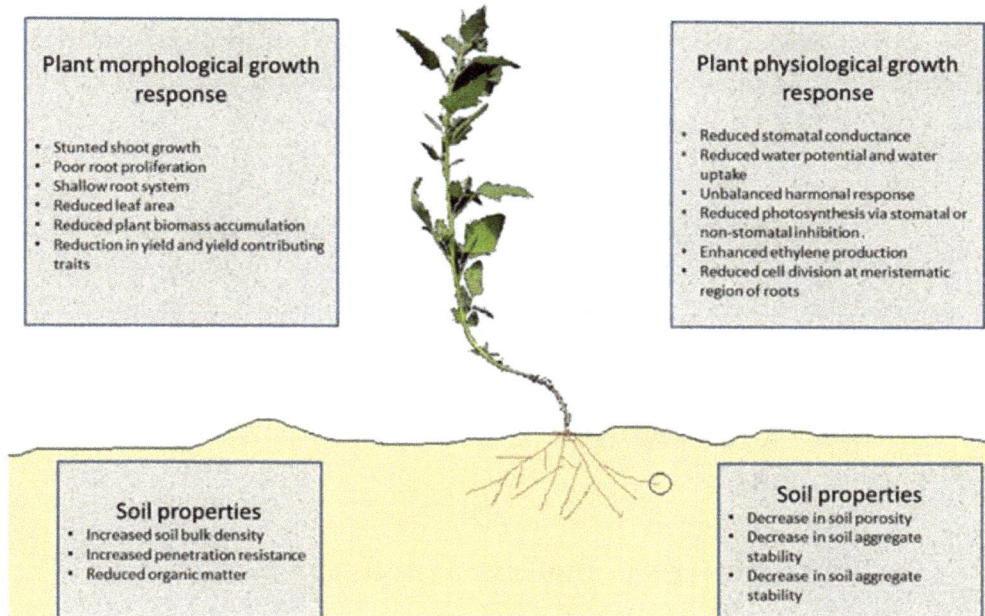

Plant morphological growth response

- Stunted shoot growth
- Poor root proliferation
- Shallow root system
- Reduced leaf area
- Reduced plant biomass accumulation
- Reduction in yield and yield contributing traits

Plant physiological growth response

- Reduced stomatal conductance
- Reduced water potential and water uptake
- Unbalanced harmonal response
- Reduced photosynthesis via stomatal or non-stomatal inhibition.
- Enhanced ethylene production
- Reduced cell division at meristematic region of roots

Soil properties
- Increased soil bulk density
- Increased penetration resistance
- Reduced organic matter

Soil properties
- Decrease in soil porosity
- Decrease in soil aggregate stability
- Decrease in soil aggregate stability

FIGURE 2.6. Summary of the knowledge of the effects of soil compaction on soil plant morphological and physiological growth and soil properties [20].

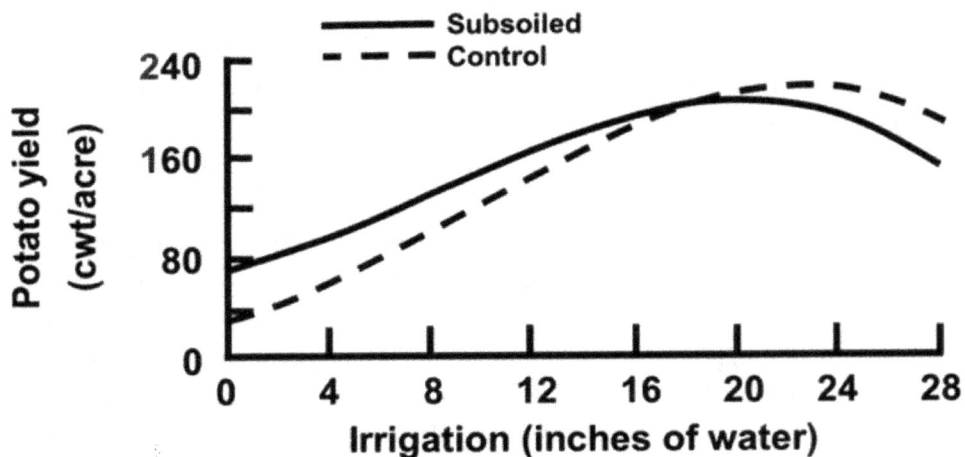

FIGURE 2.7. Potato yield at different irrigation levels for subsoil and control fields [21].

impact that soil compaction has on soil and crop health [20]. Reduction in plant growth and development, such as biomass accumulation, stomatal photosynthesis, and poor proliferation, as well as poor nutrient and water uptake decrease yield and overall crop productivity. Figure 2.7 depicts the impact on potato yield resulting from varying irrigation levels [21]. This graph shows that the availability of adequate water can increase yield by at least 100%. Because soil compaction so negatively impacts water availability and uptake, the conclusion can be drawn that compaction issues can decrease crop yield and productivity by up to 50%. In short, this also means that farmers risk losing 50% of expected profits when the soil compaction problems are not properly addressed. Inattention to this vital issue in land management can destroy a resource's ability to be productive both now and in the future. It is imperative that farm managers understand the connection between management of soil compaction today and the long-term sustainability of the agricultural ground into the future.

3. THE CAUSES OF SOIL COMPACTION

Soil compaction is the phenomenon associated with the collapse of soil media to support the loads imposed upon it. All agricultural operations on the surface of the ground cause soil compaction. Heavy axle loads, wet soil operations, livestock grazing, and materials stored directly on the surface can all result in unwanted compaction. The details of these agricultural process root causes of soil compaction will be explored in this section.

3.1 OPERATION OF EQUIPMENT WITH HEAVY AXLE LOADS

An axle load is the total load supported by a single axle, usually across two points of contact on either side of the vehicle. Although most agricultural equipment uses two axles for load distribution, each point of contact carries harmful loads into the soil. A large agricultural vehicle weighing 20 *ton* creates 10 *tons* of force on each

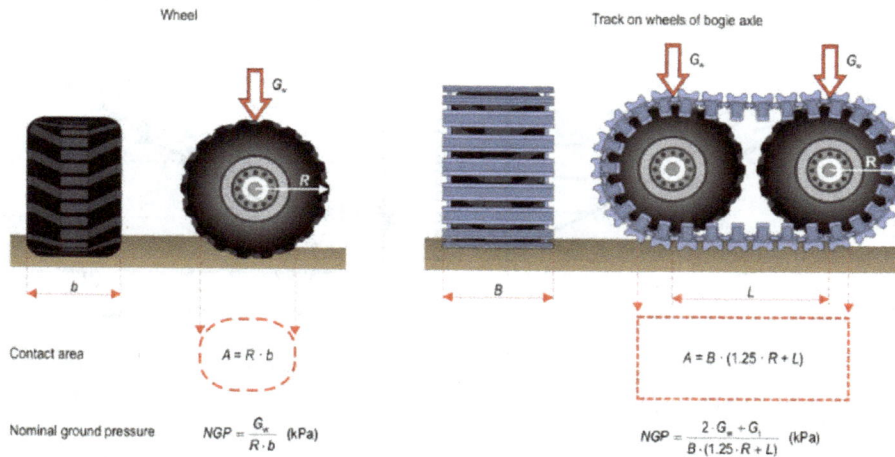

FIGURE 2.8. Tracks versus tires load distribution areas [22].

axle and causes the soil beneath each point to compact, until it can support the imposed load. The biggest factor to consider in reducing soil compaction is large axle loads. For two vehicles with the same weight distribution, the bigger the vehicle's contact area with soil, the lesser the pressure is applied to the soil surface. Figure 2.8 illustrates an advantage of tracks over tires by the contact area parameter [22]. Research has shown that having an axle load of 10 *ton* can cause deep (more than 45 *cm*) subsoil compaction under moist conditions [4]. Grain carts and other heavy trailing implements behind the power units add to the problem of soil compaction, since axle load is determined by the total weight of the vehicle divided by the number of axles. Reducing single-axle loads below five ton or less will diminish subsoil compaction and only cause topsoil compaction [4]. Using heavy machinery under wet or moist conditions always increases soil compaction dramatically over operations under dry conditions for most soil types [23]. The relationship among pressure applied, water content, and bulk density varies across different soil types as particles rearrange with changing water contents [24].

3.2 OPERATION DURING NON-OMPTIMAL SOIL CONDITIONS

Under non-optimal soil conditions, field farm operations should be considered with great reluctance due to the potential for severe damage to the soil matrix. As farm equipment crosses through a wet field, ruts are formed from soil compaction around the tire path. Tillage is a common practice to relieve soil compaction due to poor soil management. However, tilling breaks apart the soil structure and causes further traffic, in addition to deeper compaction in the field. A tilled soil is more easily compacted, since the subsoil beneath the tillage line is now in a more vulnerable state for soil compaction [25]. Under good soil conditions, the integrity of the soil is reasonably strong and minimizes the loss of pore space from heavy equipment travel. When soil conditions are non-optimal, the structural integrity of the soil is significantly reduced, and this results in the elimination of pore space with vehicle traffic. As shown in Figure 2.9, when the same pressure is applied in a loam soil, the bulk density significantly increases with increasing soil water content, thus leaving the soil

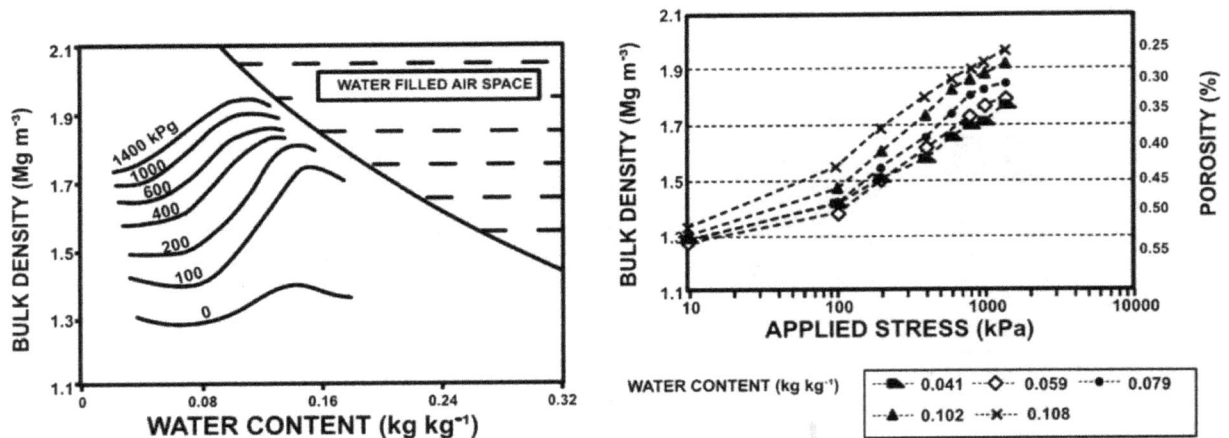

FIGURE 2.9. Water content, pressure applied, and bulk density diagram (left) and compression curve for a loam-typic Haplaquept soil (right) [24].

susceptible to compaction [24]. Additionally, water within the soil matrix reduces the coefficient of friction between neighboring soil particles and promotes the ease of displacement and flowability of the soil.

3.3 LIVESTOCK GRAZING

Livestock grazing can affect soil stability and functionality if not managed properly. The severity of soil damage due to livestock grazing is related to soil type, texture, and moisture content. Pugging, the formation of soil around the hoof of the livestock, can result in increased soil compaction and a reduction in soil surface water infiltration rates [26]. When water doesn't infiltrate through the soil surface during rainfall or irrigation, puddling occurs in fields. The trampling and pugging from livestock onto soil surfaces damages the subsurface soil integrity. The density of the livestock per unit of area in a pasture impacts the level of soil compaction due to pugging. This effect also negates the value of winter grazing on cropland to glean harvest losses. The long-term damage from soil compaction to the crop ground greatly outweighs the value of the "free" feed gained.

3.4 OTHER FACTORS

Aside from intensive farming and grazing practices common in modern agriculture, there are other factors, some environmental and some man-made, that can have a noticeable effect on soil compaction. Depending on the region of agricultural production, the type of soils, and natural and artificial drainage, some fields can be subject to prolonged ponding of water in localized areas. Over time, the weight of the water ponded on the soil surface causes the soil pores to collapse, further slowing the movement of water through the soil and increasing the weight of water on top of the soil surface during future precipitation events. Water ponded on the soil surface adds 10 *kPa* of pressure per *meter* of depth. Additionally, slowed water movement through the soil increases the risk of farming operations occurring during non-optimal soil conditions. Another non-conventional contribution to soil compaction is the relatively new practice of storing grain in large plastic bags that are laid out on the soil surface. Producers using this method of temporary grain storage have noted significant soil compaction on the surface due to the weight of the grain.

4. OFF-ROAD VEHICLE DESIGNS FOR SOIL COMPACTION MANAGEMENT

Agricultural tractive power units are the largest source of unwanted soil compaction today. Significant research and financial investment have been made in methodologies to reduce the compaction from these vehicles. Tracked systems and advanced tire systems are both designed to spread the loads imposed on the soil below detrimental levels. This section will review these common undercarriage systems, along with advanced compaction reduction technologies for off-road vehicles.

4.1 TRACKS

Commercially successful track-type vehicles, which were recognized under the trademark name Caterpillar, began production in the early 1900s [27]. These early agricultural tractors, similar to the one shown in Figure 2.10, paved the way for future tracked vehicles and the continued use of the more complex metal grouser-style

FIGURE 2.10. Benjamin Holt testing the first prototype gasoline-powered track-type tractor in 1908 [28].

tracks on construction equipment. Tracks did not remain popular in the agricultural sector once pneumatic tires became available and faded from use for many decades due to some specific issues that later rubber-belted machines were finally able to address.

While construction equipment is traditionally shipped to a worksite on a large trailer, tractors are generally driven from field to field on the road. Track-type machines with metal grousers are slower than pneumatic-tired machines during road transport under their own power. This slower transport speed, combined with a poorer ride for the operator and higher costs, eliminated most traditional track-type tractors from the agricultural market during the 1920s and 1930s. Two revolutionary designs, which are still produced by major manufacturers today, reintroduced the use of tracks on tractors. In 1986, Caterpillar launched the revolutionary rubber track Challenger 65® tractor, shown in Figure 2.11 [3]. The Challenger used a two-track running gear system, similar to most construction equipment designs [28]. Shortly thereafter, Case IH introduced an articulated tractor with tracks at each corner of the machine. The Quadtrac®, shown in Figure 2.12, was configured like a traditional four-wheel drive tractor, with each of the contact points supported using a triangular track drive and bogie mechanism [29].

Today, rubber-belted tracks have become so successful that a common argument in the agricultural world is the debate of tracks versus tires. However, opting for a tracked configuration creates a sizable increase in

FIGURE 2.11. A 1986 Caterpillar Challenger 65® rubber-tracked tractor [3].

FIGURE 2.12. A 1997 Case IH Steiger Quadtrac® tractor [29].

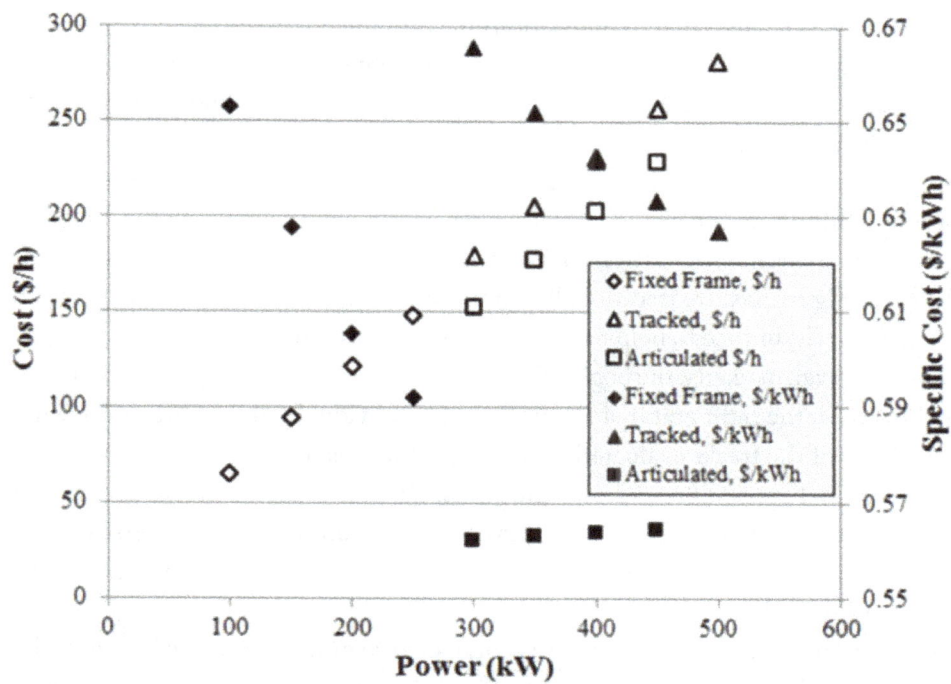

FIGURE 2.13. Costs of operating tractors on tires and tracks [30].

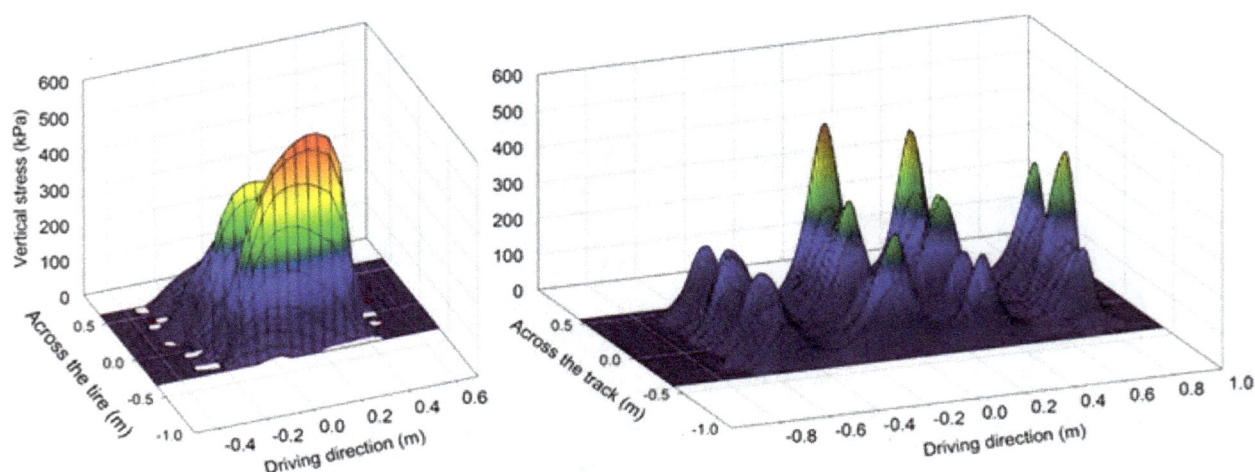

FIGURE 2.14. Soil compaction study findings for a beat harvesting machine on tires (left) and the same machine on tracks (right) [31].

both purchasing and operating costs for the tractor. The price jump from tires to tracks can often be in the neighborhood of 10–25% of the cost of the machine. The operating costs jump too. As can be seen in Figure 2.13, the specific operational cost difference between tracks and tires for a 358 *kW* tractor is approximately $0.085/*kWh* [29]. Currently, available data and existing published studies seem to support both sides of the tracks-versus-tires debate.

A 2018 European study involving a comparison of tracks and tires on two identical sugar beet harvesters revealed that the use of tracks does have a positive impact on reducing soil compaction [31]. Stress transducers were placed under the soil to analyze the compaction effects of the tractive devices. The mean ground pressure for the tire undercarriage system was measured to be 107 *kPa*, while the rubber tracks had a mean ground pressure of 84 *kPa*. As shown in Figure 2.14, ground pressure for the tires was also more concentrated and was more distributed under the tracks [31].

Because of their larger footprint, rubber tracks are often assumed to have a uniform weight distribution, but this is not true. Multiple design elements in a rubber track system, along with the integration of the track system onto the vehicle's frame, are critical to its effectiveness at reducing soil compaction. Common track systems, as shown in Figure 2.15, are traditionally composed of large driver and end wheels and smaller bogie wheels [29]. Bogie wheels, in theory, help to distribute half the axle weight across the track's contact surface with the ground. However, in reality, the bogie wheels create ground pressure spikes beneath their relative positions. As can be seen in the right graph of Figure 2.14, the individual soil pressure peaks can be attributed to the bogies and wheels of the tracks evaluated in the study. The ideal performance of a track can be identified by finding the theoretical applied ground pressure stress. This calculation is performed by dividing the load on the track by its contact area. In the sugar beet harvester study comparing tracks versus tires, the researchers discovered that the peak stress applied to the ground by the tracks was 5.7 times greater than the ideal ground pressure calculated value [31].

An analysis of the soil types across Europe was conducted to evaluate the maximum load capacity of different tractive devices without causing permanent soil deformation [31]. Figures 2.16–2.18 convey this analysis, showing soil types and the respective loads that can be handled by tires, tracks, and ideal tracks having a

FIGURE 2.15. Bogie wheel track design [29].

uniform pressure distribution. It is worth noting that a substantial load-bearing increase could be achieved through improved track design.

Regardless of which side of the track-versus-tires argument is seen as the correct economic option, tracks do serve utility for farmers beyond that of tires. Although farmers prefer to be in the field when conditions are good, the weather does pose challenges. Depending on the geographic location of a farm and its soil type, it is common to deal with wet field conditions. As a result of the need to beat seasonal weather patterns, farmers often push acceptable limits to finish critical tasks in the field. Saturated soils are easier to tackle with tracks because of their improved tractive performance over tires. Tracks are also less prone to rutting the soil in wet conditions. As seen in Figure 2.12, a side-by-side comparison of a tracked and a tired machine shows the improved performance of tracks at staying on top of the soil. As can be seen, the track's footprint barely marks the ground, whereas the trailing tire cuts a deep rut. Severe soil compaction, such as that caused by the tires in Figure 2.19, negatively impacts the health and performance of a farm's soil for the long-term [32]. However, the inability to harvest crops negatively impacts the economics and viability of a farm's business today. Although tracks are pricier than tires and may only provide limited benefits toward economically reducing soil compaction, tracks clearly outperform tires in adverse conditions.

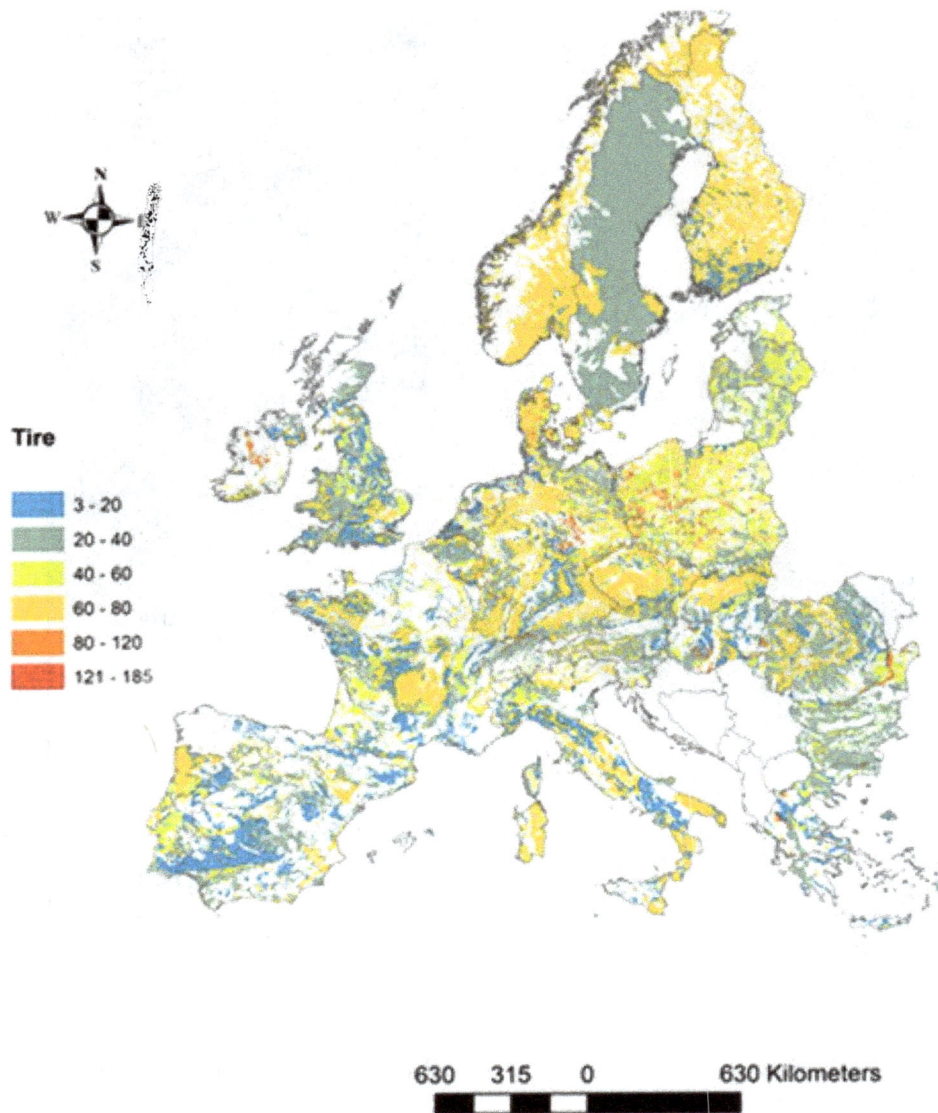

FIGURE 2.16. European soil load-carrying capacity map, showing the maximum load (*kN*) that can be carried by a 1050/50R32 tire without inducing permanent soil deformation at 0.35 *m* depth [31]. The lowest load-carrying capacity exists in the United Kingdom and southern portions of the continent; the highest in Scandinavia and the northeastern portions of the continent.

4.2 LOW-INFLATION TIRES

Although the European study concluded that the use of tracks had a positive impact on soil compaction, differing studies have led to opposite conclusions [31]. The argument for tires is that correct maintenance needs to be performed to ensure that the tires are inflated correctly. A common issue is that farmers will over-inflate tires. Studies such as the one highlighted in Figure 2.20 reveal that incorrectly inflated tires create the most compaction [33]. In this specific experiment, correctly inflated dual tires were found to be impressively less compacting to the soil than tracks while demonstrating that tracks could be superior to poorly maintained dual tires.

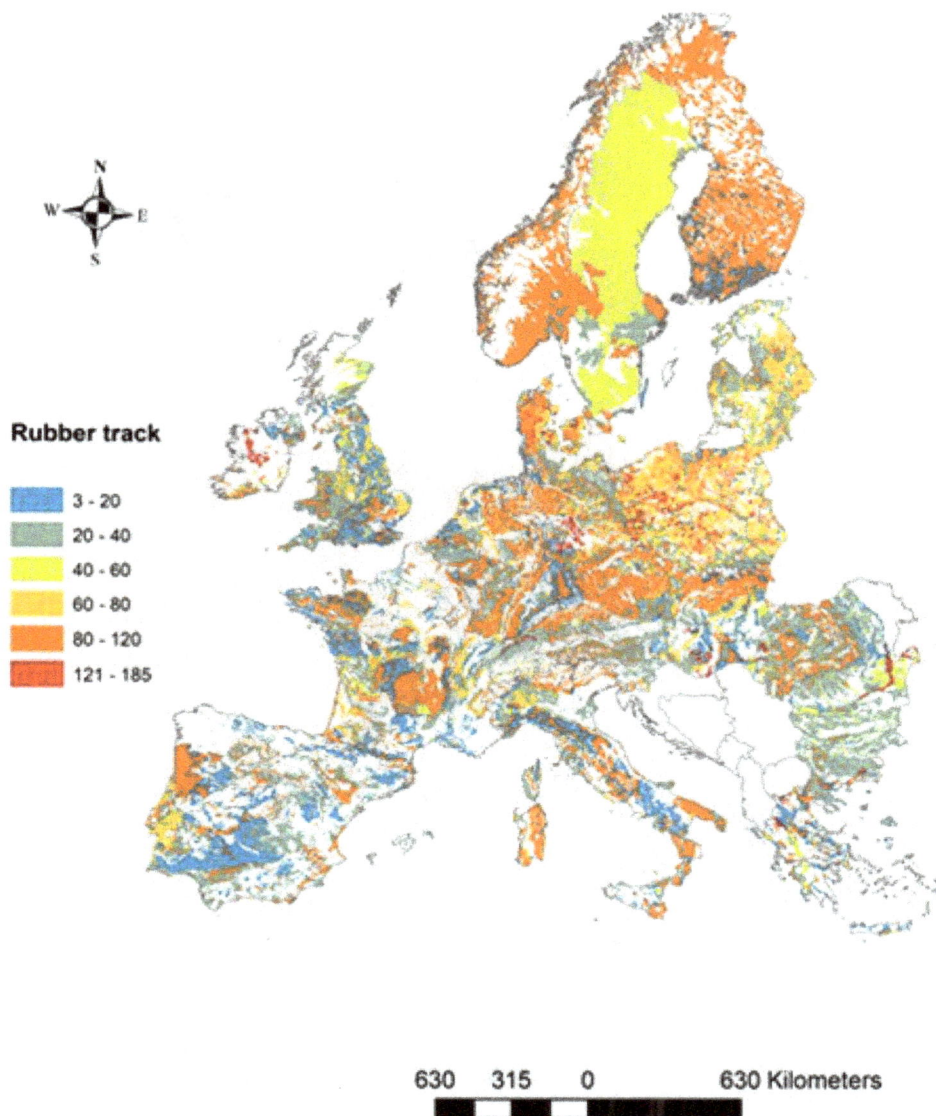

FIGURE 2.17. European soil load-carrying capacity map, showing the maximum load (*kN*) that can be carried by a rubber track without inducing permanent soil deformation at 0.35 *m* depth [31]. The lowest load-carrying capacity exists in the United Kingdom and southern portions of the continent; the highest in Scandinavia and the northeastern portions of the continent.

Tracks are undeniably an expensive but great option for reducing soil compaction. However, low-pressure, properly inflated tires are a potential option to match the benefits of tracks at a fraction of the price. It is well documented that the depth of soil compaction is strongly correlated with tire pressure. Lower pressures cause less compaction. The limit to this practical method of reducing compaction is that the bead of the driving tires must remain on the rim. This low-pressure tire strategy helps reduce soil compaction by increasing the tire surface contact area with the ground. The increased contact area reduces the pressure exerted on the ground. Due to the limitations of decreasing the air pressure in traditional radial tires, tire companies have developed new flexion technology to allow even lower tire pressures. Increased flexion and very increased flexion tires, first introduced by Michelin during the 2000s, use a mature technology that greatly decreases the soil

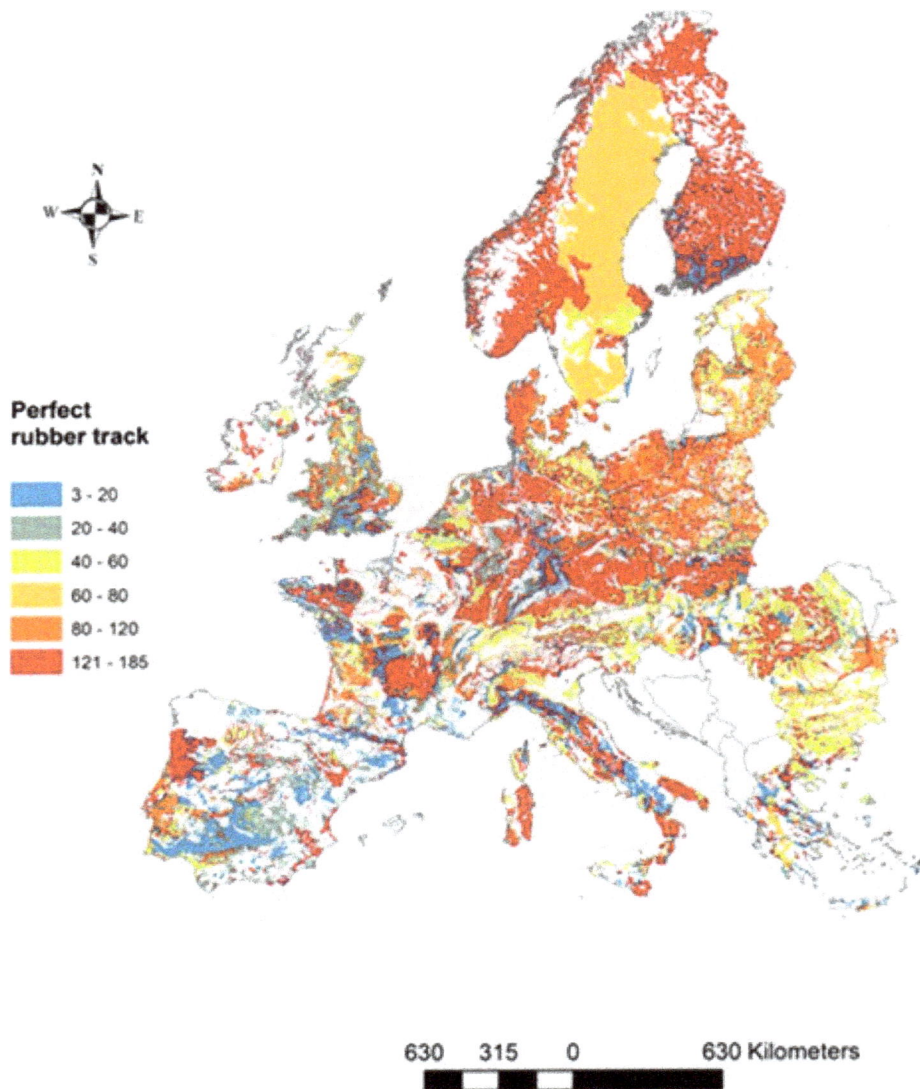

FIGURE 2.18. European soil load-carrying capacity map, showing the maximum load (*kN*) that can be carried by a rubber track with perfectly even stress distribution without inducing permanent soil deformation at 0.35 *m* depth [31].

compaction from today's heavy machinery. The very increased flexion and increased flexion tires can support the same loads with 40% and 20% less air pressure, respectively, than radial tires by using increased tire sidewall strength [34].

While tires do not have as extensive a surface area as most track designs, low pressure and flexion-style tires make up for some of the ground pressure shortcomings on tired vehicles and in some applications can be a more viable option. Tracked vehicles experience pressure spikes at each bogie, whereas tires can be more consistent in the application of load to the soil [35]. Modern row crop tractors are commonly seen with dual rear tires and even dual front tires. As new equipment becomes larger and larger, single tires are no longer viable. As the demand for modified front-wheel drive (MFWD) tractors has increased, additional weight has been

FIGURE 2.19. Rut comparison of tracks versus tires in muddy conditions [32].

added to the machines, requiring further soil compaction reduction methodologies to be undertaken. The most common MFWD tractor variants today have both dual front and rear tires. As would be expected with the addition of a second set of tires, soil compaction is reduced. This is achieved by essentially doubling the contact surface area [36]. The addition of a second set of tires allows for tire pressure to be reduced even further, also decreasing the potential for soil compaction [36]. These strategies can be combined for reasonably additive results. Using flexion-type dual tires at low tire pressures can achieve even lower soil compaction. Under certain circumstances, properly inflated duals have been shown to be more effective at reducing soil compaction than tracks. Triple tires can be seen in certain high-power applications. However, they are not commonly seen in modern agriculture. The increased width of the tractor would be a benefit in the field, but transport on the road becomes infinitely more challenging. Axle stresses multiply significantly as well.

A recent innovation in agricultural tractor tires involves changing the overall design of the tires and the rim. New low sidewall (LSW) tires feature a significantly reduced tire aspect ratio, which results in a wider tire with reduced sidewalls. These LSW tires are intended to completely replace duals on modern farm equipment. While these new tires are a more expensive initial investment, including a completely different rim and new tires, they are still cheaper than modern tracked systems. LSW tires could be a viable option for reducing soil compaction as well as providing other benefits to the operator [37]. Tractors with a high center of gravity and LSW tires can experience reduced sway in motion as well as better resistance to power hop [37]. LSW tires have a larger width, allowing for more surface contact with the soil as well as retaining the reduced inflation pressures similar to the flexion-style tires [37].

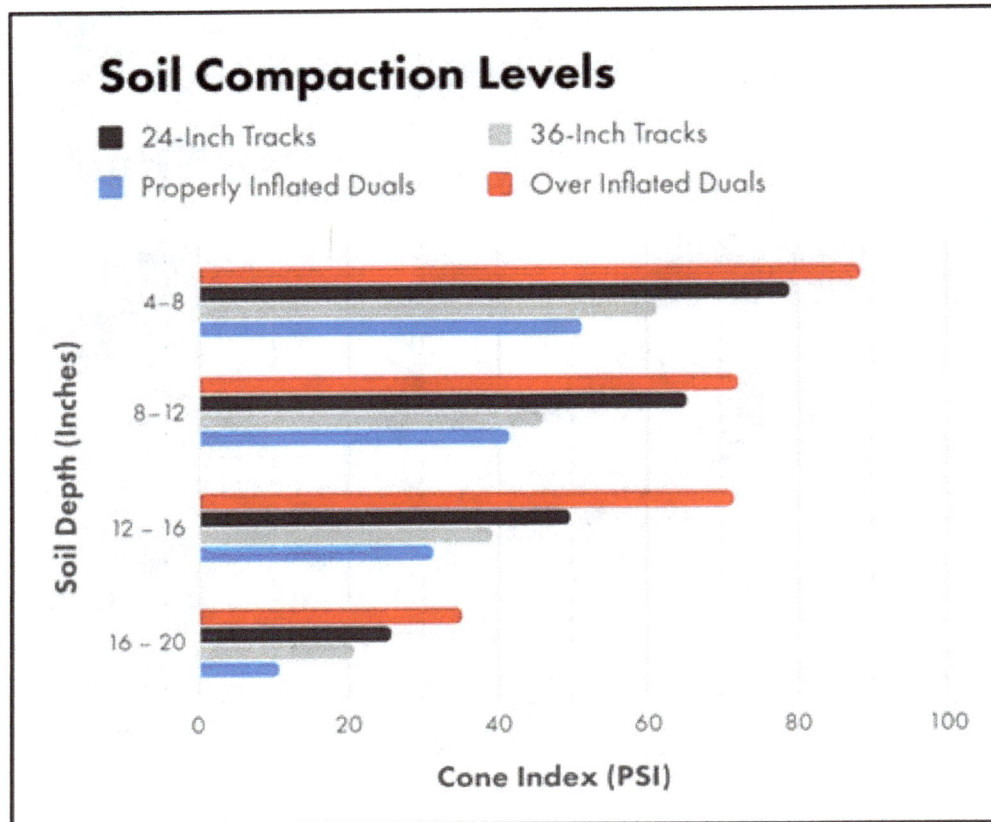

FIGURE 2.20. Soil compaction comparison study findings [33]. At all depths, the over inflated duals correlate to the highest cone index, followed by 24-inch tracks, 36-inch tracks, and finally, properly inflated duals.

Tires are typically the more attractive option for most farms due to lower purchase and operational costs. Since the potential benefit of tracks is only for the reduction of soil compaction, tire systems, which can effectively compete with tracks in this metric, have a competitive advantage. With LSW tires closing the marginal difference in performance between tracks and tires, tires in many standard applications may be the smarter option. However, any option to reduce soil compaction will pay out in the long run for farmers and growers. Conservation of the world's natural resources is imperative for the continued survival of humanity, especially with the extreme population growth projected for the next 50 years. Producing more food with less resource inputs is the goal of all of agriculture. Conserving the land is the first step toward a better tomorrow that will continue to be able to feed people from the soil.

4.3 TWO- VERSUS FOUR-WHEEL DRIVES

Four-wheel drive vehicles can produce less soil compaction than their two-wheel drive counterparts, assuming that all other factors are equal. Four-wheel drive systems also encounter less slip in motion and have a more optimal weight distribution, which likewise helps reduce soil compaction. Slip can be thought of as a horizontal component of soil compaction. Slip occurs as the soil behind the tire compacts to support the drawbar load. There is a shrinkage in the matrix of the soil [38]. When traveling off-road, all vehicles have some

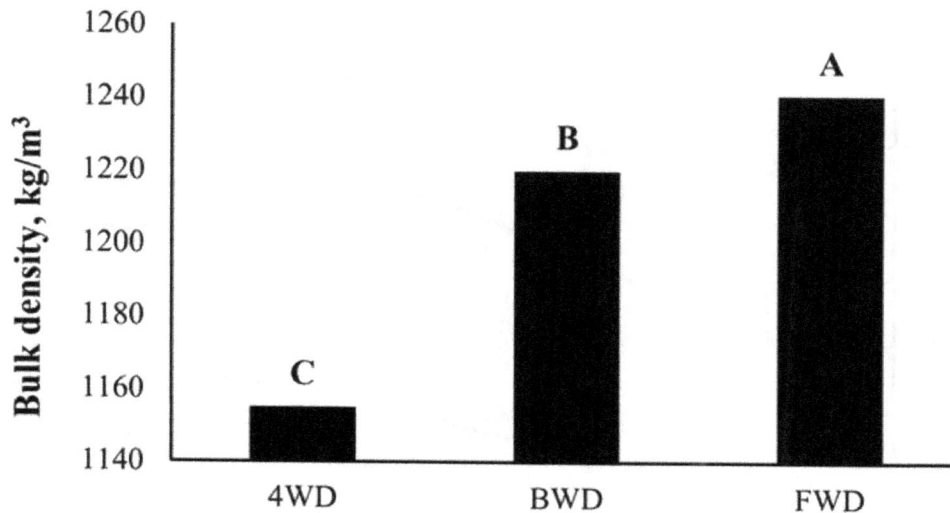

FIGURE 2.21. Bulk density of soil following a tractor pass with different drive systems [40].

amount of slip. This slip is determined by the interface between the wheels and the ground. The larger the ground contact area, the less slip occurs. Four-wheel drive vehicles have less slip than two-wheel drive vehicles. While two-wheel drive vehicles may have the same number of wheels on the ground, the non-driving wheels do not provide any traction. Tracked vehicles have an advantage because the entire length of the tracks are driven and therefore have reduced slip when compared to tires. Nonetheless, slip sufficient to support the forward travel and drawbar loads on the machine still occurs. The reduction in soil compaction behind four-wheel drive vehicles has been demonstrated experimentally. Figure 2.21 shows that the bulk density of soil was found to be 5.6% less than in rear-wheel drive and 7.3% less than in front-wheel drive vehicles [39].

From a practical perspective, four-wheel drive and two-wheel drive vehicles are built differently. Four-wheel drive vehicles are designed to have a different weight distribution. A rear-wheel drive vehicle has the center of mass at roughly one-third of the wheel base forward of the rear axle. A four-wheel drive vehicle has the center of mass located slightly more forward. This is advantageous, because the tractive force from the wheel depends on the normal force with the ground. Under drawbar load, the front and rear ends of the tractor are supported more equally, and the peak pressure on the ground is lower. Larger wheels have a higher area of contact with the ground, which results in lowered soil compaction. Many four-wheel drive vehicles have an articulated chassis used for steering. An articulated vehicle's axles follow only a single pathway when turning, which also reduces the area of compaction.

Just as certain soils are more prone to soil compaction, some soil types benefit more from four-wheel drive tractors. It is more difficult to gain traction in loose soil. As Figure 2.22 shows, the moisture content in the soil plays a significant role in the compaction tendency of the soil [39]. Soils with a greater moisture content typically generate more slip [41]. As discussed earlier, slip is correlated with soil compaction. "Tire travel" will be significantly more in wet soil to cover the same distance.

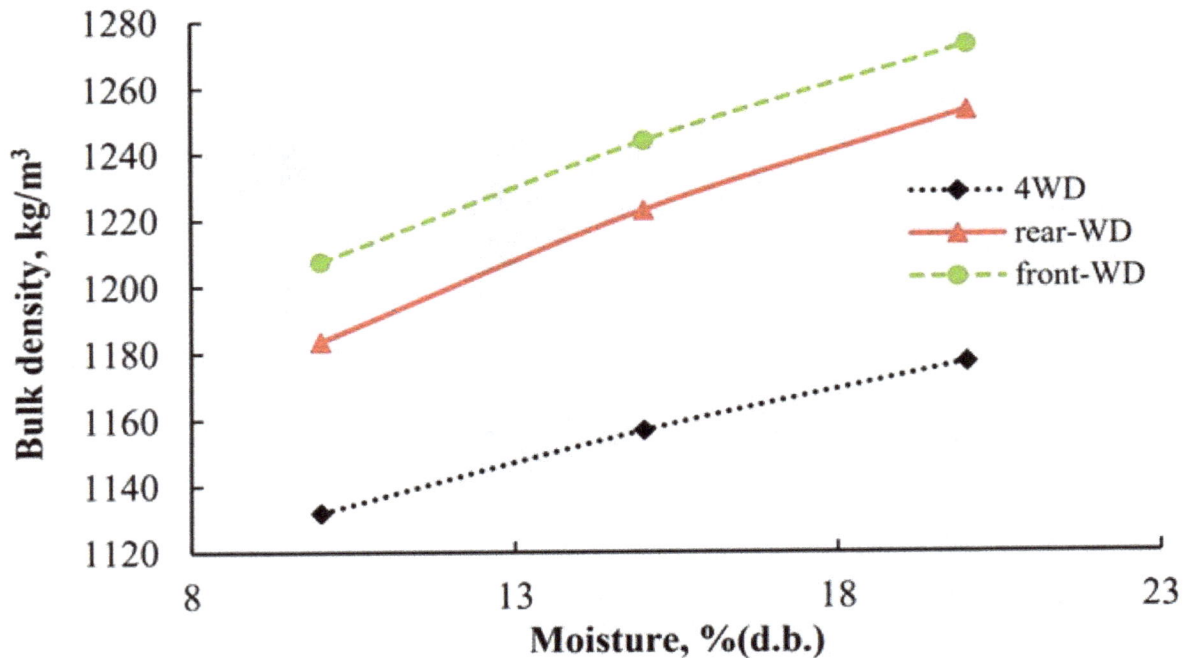

FIGURE 2.22. Effects of moisture on soil compaction between multiple vehicles [40].

4.4 SENSORS, ACTUATORS, AND SPECIAL APPLICATIONS

Mechatronic agricultural systems are the future of agricultural machinery. One proposed means to reduce soil compaction is to utilize numerous smaller robotic machines instead of progressively larger machines to tend the fields. One limitation to further development along these lines is the price of a fleet of machines, while another is the human management factor. The price will likely come down as the technologies develop, but the human factor will remain stagnant until a critical mass of the new equipment enters the agricultural equipment market and demonstrates viability. These modern agricultural mechatronic systems will contain numerous sensors and actuators as their essential elements. Actuators perform the specific tasks directed by the vehicle's controller. Sensors facilitate the feedback from the actuators to the tractor's control system. The feedback data works as a performance measure for the actuators and the control system as whole. Specifically, the control system receives data on the success of the actuators' actions, the vehicles' position and motion, and the vehicle's immediate environment. Driveline control systems with feedback mechanisms can successfully address soil compaction problems in many special applications. These automatic control systems have potential for use in the envisioned swarm systems for everyday agricultural operations. A swarm of smaller automated machines could become a disruptive technology, which would shift the paradigm in the current soil compaction reduction practices for crop production systems. The utilization of real-time feedback from soil conditions has multiple previous implementations for experience to be drawn from.

One example is a unique tractor for special climates. The Gidrokhod 49061 ("Gidro"—hydro "khod"—traveler), shown in Figure 2.23, is "a three-axle all-wheel drive machine with a hydrostatic driveline with an

FIGURE 2.23. Gidrokhod 49061 [42].

automatic control system" [42, p. 147], which combines an individually driven axle design with a feedback-based approach to vehicle control. The Gidrokhod's driveline operates as follows: "[The] driveline is a full-flow mechanism that includes three axial-plunger controllable reversible and invertible hydraulic pumps and six axial-piston controllable and invertible hydraulic motors. Each pump is associated with two tandem hydraulic motors that set into motion the wheels of one hypothetical axle. The torques and rotational speed of the hydraulic motors are controlled individually by varying the displacements of the pumps and motors by means of an automatic control system" [42, p. 147]. The Gidrokhod's automatic control system supplies the required power to each wheel "as a function of the current conditions of interaction with the soil" [42, p. 147]. Gidrokhod was originally designed to reduce soil compaction from human activity in tundra, where the plants and soil are particularly vulnerable to any soil loading, such as the pressure from tracks or tires. The Gidrokhod's hydrostatic driveline also improves the vehicle's off-road drivability by dampening returned ground shocks into the driveline. The Gidrokhod's hydrostatic driveline is a computer-controllable tested technology that could be transferred to off-road vehicle applications in agriculture to address the soil compaction problem.

Another example of a special off-road vehicle is a small off-world exploratory rover. These machines closely resemble hypothetical swarm agricultural vehicles and are essentially miniature space tractors. The pace of modern technology suggests that humanity will start colonizing the Moon and Mars by the mid-21st century. The comparison between an off-world exploratory robot and a small swarm agricultural robot tractor is not as outlandish as it might first appear. The sensory apparatus necessary for independent wheel suspension control works as well for minimizing soil compaction as navigating unknown terrain. The external manipulators resemble plant-tending tools, and the ability to remain on station and function unmanned is similar. It is conceivable that low-compaction–inducing swarm agricultural tractors may look a great deal like our exploratory robots.

Russia was the first country to send a rover to the Moon and Mars. While the Mars mission was a failure, the Soviet Moon exploration program laid the foundation for the future robotized space exploration missions. Over 50 years ago, in 1970, the Soviet Lunokhod 1 ("Luna"), shown in Figure 2.24, had a number of soil compaction sensors, a special wheel to measure traction, and single independent drives on each of its eight wheels for improved mobility [43]. As of now, multiple Mars rovers from the United States and one from China are traveling on the surface of Mars. Modern US rovers such as Perseverance, shown in Figure 2.25 [44], combine the essential soil compaction sensors with the sophisticated modern drivetrain solutions, such as an advanced feedback loop from a complex sensor network, photo and video surveillance systems, and the use of Big Data concepts to better predict the ambient soil conditions and any possible action protocols during deployment and moving between operation areas.

The off-world researchers operating these rovers build terrain models on Earth using transmitted data from the operating rovers, which has a 5- to 20-*min* signal delay. They use location information from satellites circling Mars, just like farmers on Earth do for agricultural production. Similar to how military location technologies came into the consumer world, space-based technologies will eventually find their best-use applications

FIGURE 2.24. Lunokhod 1 Moon rover [43].

FIGURE 2.25. Perseverance Mars rover [44].

on Earth. One particular technology transfer path will be for the highly accurate location technologies needed to control small robotic low soil compaction-inducing vehicles for agricultural production. The first half of the 21st century will continue to see increasing technology transfer from electronics and space industry into everyday agricultural operations.

5. SOIL COMPACTION MANAGEMENT IN AGRICULTURE

Although the chassis and undercarriage design of tractors, combines, and harvesters is of obvious concern to engineers when trying to decrease soil compaction, agriculturalists have developed a variety of other methods and practices that contribute to the alleviation of soil compaction impact. These conservation tillage practices and alternative process design considerations are an important element in overall soil compaction reduction, as they can be applied to any and all farming operations even those that do not have the most up-to-date equipment. Farm management practices can have a profound impact on reducing soil compaction as well as maintaining soil organic content, reducing nutrient suppression, and decreasing time and energy spent in the field. From a sustainability standpoint, these conservation practices may even be more impactful for farm and field management at reducing soil compaction than any specific tractor or undercarriage design.

5.1 TILLAGE EQUIPMENT AND PRACTICES

The design of tillage equipment is an important and fruitful area of research for reducing soil compaction during necessary ground-working operations. Tillage equipment design affects the ways in which tillage equipment interacts with the soil to help alleviate long-term effects of disturbance in heavily worked ground. Some of the core ideas in tillage implement design are load distribution, working point and shank design, working depth, the different types of soil disturbance, and the soil pulverization level. Most modern research is targeted at collecting specific information about the impacts of these conditions on soil health, compaction levels, and seedbed preparation as well as energy use and the time spent in the field. The implications of tool design, structural loading, types of conservation and reclamation equipment, and impacts of soil compaction management on energy consumption and overall performance of an agricultural venture will be examined in this section.

5.2 TOOL DESIGN

In early agricultural practices, the moldboard plow dominated tillage as the most effective tool for turning the soil to create a seedbed. Its design was maintained in many forms of tillage equipment for years before its harmful impact on soil health, organic material, and erosion was realized. In contrast to the simplistic design of the moldboard plow, modern tillage equipment tool designs come in many shapes and sizes. The effects of tool geometry, orientation, depth, field speed, and other factors impact the level of soil disturbance and compaction. Various tool types can create a multitude of different outcomes in the upper soil layers in terms of soil aggregate size, topsoil density, porosity, and organic matter distribution. Other tools act predominately at the subsurface level. In particular, deep-cutting tines have the greatest impact on subsoil compaction. Their shape, working depth, and spacing all affect resulting soil compaction differently.

The effects of specific tine geometry and individual tine orientation were explored by researchers using finite element analysis (FEA) modeling [45]. Figure 2.26 depicts the range of geometric variation explored, including the alteration of tine width, rake angle, and tilt angle. The primary results from this study concluded that at comparable field speeds, the increase in tine width linearly increased the resulting downward vertical force, while increasing rake or tilt angle linearly decreased the downward vertical force [45]. The implications of this study affect tractor power sizing, the uniformity of transmitted force along a vertical soil profile, the soil pulverization level, and the subsoil compaction. Most certainly, the results also present a variety of design trade-offs depending on the immediate and long-term priorities of the specific farm manager. However, from the standpoint of reducing compaction whilst maximizing surface soil pulverization, minimizing the tine width and maximizing the tilt and rake angles create the least amount of subsurface compaction.

It is important to note that the tilt angles can be non-uniform on individual tines and on the overall tine setup for an entire tillage unit. Many times, a compromise between minimizing draft forces, decreasing compaction, and managing soil upheaval can be achieved by applying a diverse range of different geometric and orientation values throughout a single tillage implement [46]. This becomes even more applicable the larger the implement is due to the increased number of rows and columns of working points. Besides the considerations outlined above, two other vital components of tillage implement design are tool spacing and working depth in relation to critical depth. Critical depth is generally considered to be the point below which soil disturbances are concentrated near the working point and not distributed throughout the soil. Figure 2.27 shows both the effects from operating below a critical depth and the dramatic increase in soil compaction as a result of tillage below this level [46].

Unfortunately, critical depth is not uniform by any means. It varies significantly with multiple variables and can be heavily impacted by moisture level, soil type, and the presence of a cover crop. This makes determining an operational depth a challenging task, particularly for inexperienced operators. Initial passes are often needed

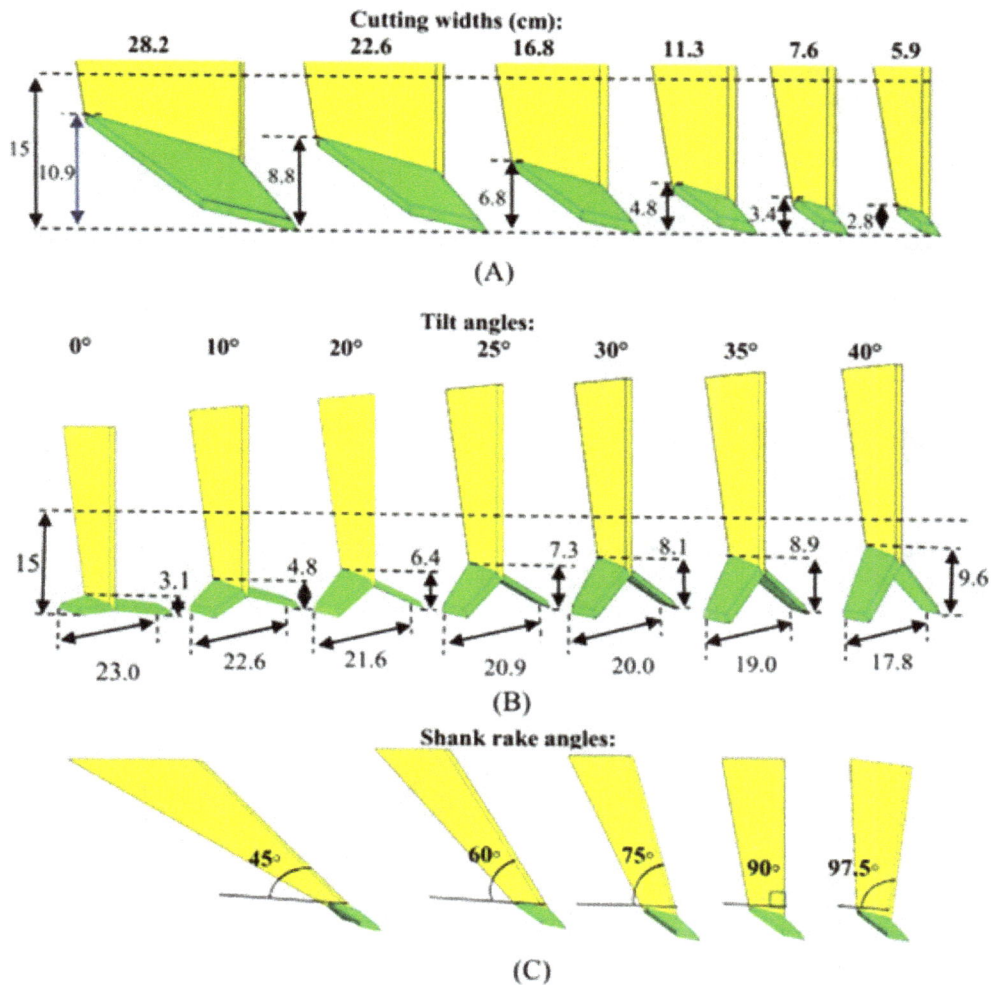

FIGURE 2.26. (A) Six single sideway-share subsurface tillage implements with the same rake and tilt angles of 10° and 15° with different cutting widths, (B) dual sideways-share subsurface tillage implements with rake angle of 15° with different tilt angles, and (C) five dual sideway-share subsurface tillage implements with share tilt and rake angles of 10° and 15° with different shank rake angles [45].

to estimate ideal working depths. There has been some research regarding the use of strain gauges on subsoiler tines in conjunction with depth sensors, which can utilize a closed loop response system automatically adjusting height to maintain the desired draft and vertical forces [47]. These systems still require a degree of experience and skill to determine the expected shank loading at, above, and below the critical depth in order to set the necessary system limits prior to operation. Although the practical difficulties with feedback-based systems are numerous, increased implementation of the depth adjustment mechanisms described above will provide a wealth of data regarding forces at and around critical depth. This information will only make these systems more effective in the future [47]. Figure 2.28 illustrates the effects of tine spacing on overall soil disturbance. When tine spacing exceeds 1.5 to 2.0 times the working depth, an interesting phenomenon takes place in which the soil disturbance only occurs locally and results in a non-uniform subsoil profile and soil surface [46].

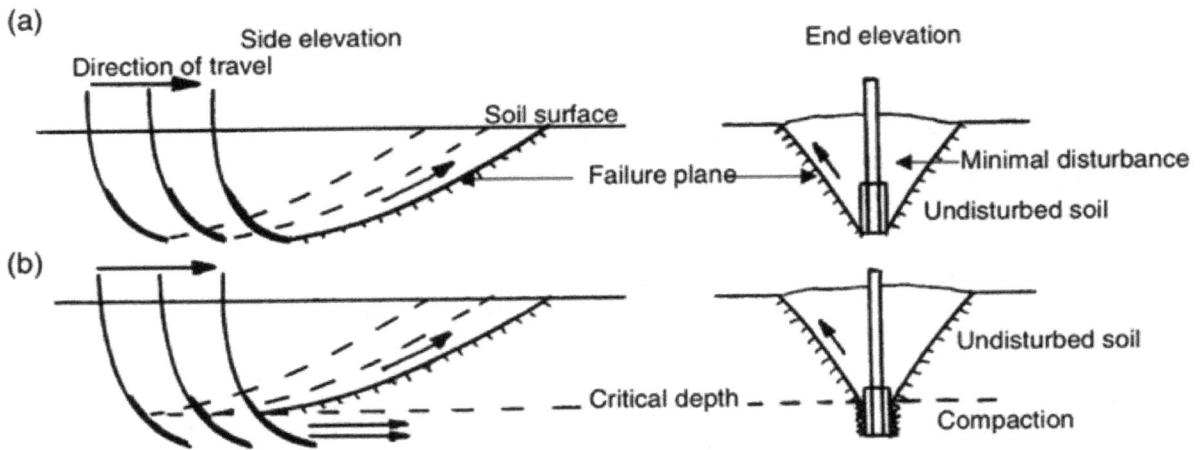

FIGURE 2.27. Varying level of soil disturbance with narrow tine: (a) Above critical depth and (b) below critical depth [46].

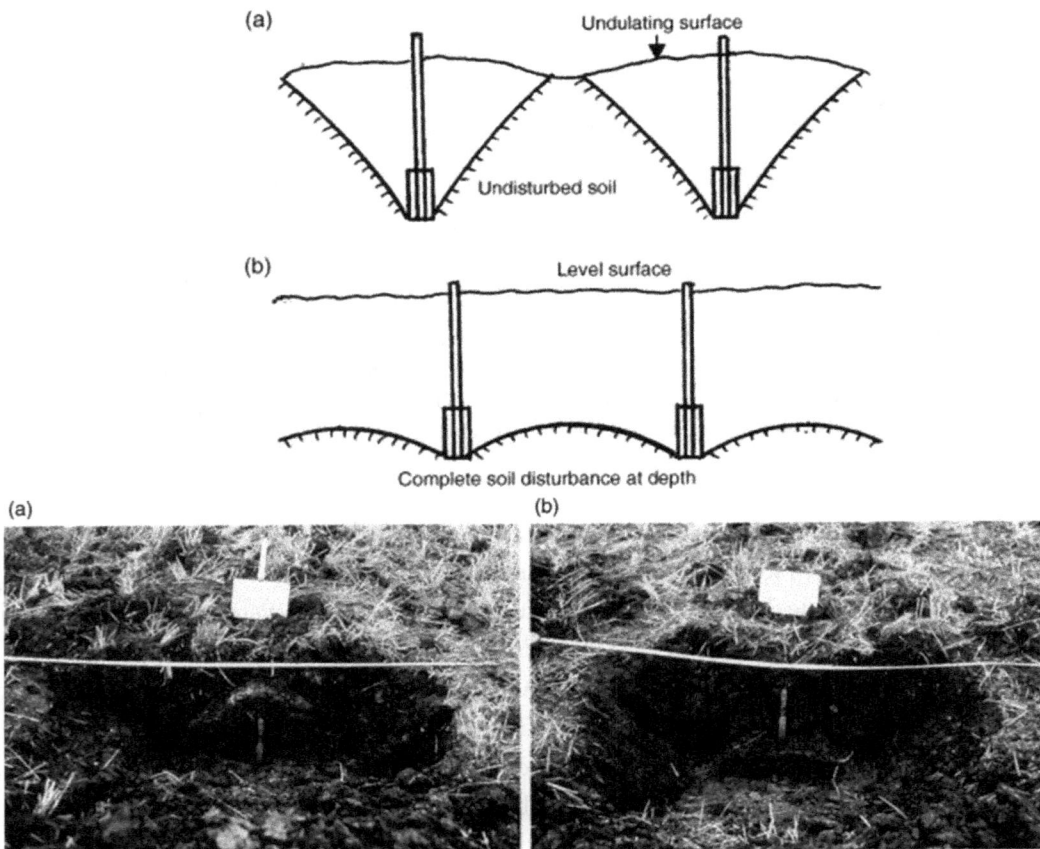

FIGURE 2.28. Influence of tine spacing on the soil disturbance profile: (a) wide spacing and (b) narrow spacing [46].

This outcome is likely to be troublesome for planting, as different row unit depth wheels will be penetrating the surface soil to different depths. The lack of consistency in seed depth, because of this poorly prepared seedbed, will result in emergence and germination issues. Although not initially obvious, the lack of uniform soil disturbance also affects compaction levels. First, the lack of a uniform soil disturbance cross-section that occurs when using widely spaced tines, illustrated in Figure 2.21 (A), results in some subsoil being undisturbed. This soil remains compacted over time. When using a tillage implement with a wider tine setup, it is easy for an operator to exceed the critical depth in order to achieve a cleaner surface profile, but in doing so the subsoil compaction has been increased throughout the field. Using a narrow tine design dramatically decreases the chances that an operator will need to exceed critical depth in order to achieve the desired seedbed quality.

5.2.1 STRUCTURAL LOADING

While magnitude of downward vertical force for tillage equipment simply does not compare to tractor units, it is still important to consider how the soil reacts with the implement loading as it moves through the field and what factors play into determining the optimal number of tines and the structure of tillage equipment. There are three primary ways in which soil reacts to the loads and forces placed on it by cultivation implement: brittle loosening disturbance, compressive disturbance, and tensile disturbance [46]. Brittle loosening occurs when the implement load compresses the soil and causes a sliding or slipping during the operation. The effects of the sliding and slipping are such that the soil aggregates, clumps, and masses move relative to one another. The overall volume of soil masses is increased, cracked, and spread out. Contrary to compressive disturbances, a large quantity of the soil is actually decompressed or loosened as a result of brittle loosening. This is the kind of soil response that occurs primarily under ideal loading and working depth conditions.

Compressive disturbance also occurs under compressive loading but without exposure to masses sliding relative to one another. In this case, without sliding, the soil is more likely to experience high degrees of compression and increases in density. This process is more common using heavier implements, when there is a low draft force. Tensile disturbance is virtually the same as brittle loosening and has similar results, such as decreased density and alleviated compaction. The difference lies in the fact that tensile disturbance occurs when soil aggregates are pulled away from one another and forced to spread out. This kind of disturbance is more likely to occur under high moisture conditions, where the load is cushioned and absorbed to a greater extent, thus negating the compressive impact of the load.

Each of the three kinds of soil matrix disturbances can be modified and impacted by the working depth, operation speed, and weight of the implement. Table 2.1 provides the basic tendencies for determining the design of the implement, based on the power of the tractor unit, and for potentially determining necessary engine power or anticipated working depth, based on the tine and structural design of the tillage implement and its working points. Table 2.1 can be used for reclamation projects in which the soil has experienced long-term compaction and where aggressive subsoiler action is needed to prepare the soil for further tillage and planting preparation [46].

5.2.2 SOIL-LOOSENING EQUIPMENT

There is a big difference between common tillage equipment used for routine crop cultivation, associated with planting and harvest, and machinery used to rejuvenate the soil from excess compaction. Robust subsoilers are utilized when efforts are made to restore long-term compacted soil. These subsoilers must be capable of decreasing soil density and effectively disturbing the mid-subsoil level to make the land workable under normal cultivation protocols. As seen in Figure 2.29, these reclamation subsoilers typically utilize a three-point

TABLE 2.1. *Wheeled tractor capability for operating loosening tines in compacted soil [46]*

TRACTOR SIZE		CAPABILITY	
ENGINE POWER (HP/KW)	BALLASTED WEIGHT (TONNES)	WORKING DEPTH RANGE (CM)	NUMBER OF TINES
30/23	1.5	20–30	1
60/45	3.0	30–40	1
75/56	3.8	35–45	1
		25–30	2
100/75	5.0	40–50	1
		30–35	2
		25–30	3
125/95	6.3	45–55	1
		35–40	2
		25–30	3
150/110	7.5	50–60	1
		35–45	2
		30–35	3
		25–30	4
200/150	10.0	40–50	2
		35–40	3
		30–35	4
		25–30	5
250/185	12.5	45–55	2
		40–45	3
		35–40	4
		30–35	5
		25–30	6

For crawler tractor in the same horsepower range, increase the number of tines by 50% or the working depth by 20%.

hitch attachment for depth adjustment rather than a drawbar attachment and trailing configuration [46]. One issue with these subsoilers is the need to operate below the critical depth to create an adequate soil disturbance to restore the soil profile. Unfortunately, this process can cause further deeper subsurface soil compaction, despite alleviating the compaction in the upper subsurface soil levels.

5.3 CONTROLLED TRAFFIC FARMING

Since compaction is inevitable in agricultural operations, its minimization through operational management is critical to long-term sustainability. The essential principles of compaction management are the reduction of both tillage and field traffic. Modern best practices decrease these elements in crop production processes to the smallest feasible levels. Compaction mitigation techniques are reviewed in this subsection.

Limit of soil disturbance

FIGURE 2.29. Reversible subsoiler and its impact [46].

5.3.1 LOW-TILL

The first management practice that can be used to reduce soil compaction is the low-tilling method. There are several aspects to low-till that help reduce erosion and soil compaction collectively. Low-till involves planting with a seed drill after a minimally disturbing tillage operation. The soil is not as exposed to and penetrated by wind and water under this protocol. Low-till keeps an estimated 30% minimum of crop residue on the soil surface. This allows for more organic material to remain present in the topsoil, increasing soil stability [48]. With low-till, water erosion is inhibited due to the higher surface trash coverage and the lower general depth of water penetration. Low-till farming protocols are currently an extremely popular choice, creating a nice compromise between conventional agricultural practices and more extreme conservation processes.

5.3.2 NO-TILL

No-till farming is extremely effective at helping soil health in multiple different ways. With this method, only the soil surrounding the seed trench is tilled by the row crop planter. No additional tillage operations are performed. Besides being extremely cost-effective in fuel consumption, no-till operations have very quick positive results compared to other methods. In as short as two to five years soil compaction will naturally be reduced in the topsoil, as microorganisms and organic material increase and expand in the soil. The increased biomass will have a longer-lasting effect, as the crushing strength of the soil will be dramatically increased. In clay soils, these results may be more pronounced. Compacted clay soils create the tightest restriction of all soil types. Allowing for root penetration and added biomass expands clay soil until it is much less compactable. With the no-till method, the higher vegetative density alone can help absorb the impact from smaller implements. With the proper planting equipment, the no-till method is a very simple and effective method for reducing and reversing soil compaction [49].

5.3.3 DEDICATED TRAMLINE EQUIPMENT

The newest realm of controlled traffic farming incorporates unified implements that minimize in-field travel in a variety of ways. NEXAT GmbH is a leader in this field. The company has developed a single equipment carrier, known as a beam tractor, capable of planting, soil cultivation, crop treatment, and harvesting. The company refers to this as the NEXAT System [50]. This fascinating piece of equipment, pictured in Figure 2.30, manages to minimize the required crop production machinery, is fully integrated, and does not require additional equipment or chassis components. It can keep up with the advancing digital age of electronic controls

FIGURE 2.30. NEXAT system for controlled traffic farming [50].

and even has autonomous guidance. However, its most impressive feature is the ability to reduce the land driven on from 60–80% to less than 5% by only traveling on dedicated drive lanes. NEXAT-like systems are crucial to the continuing effort of reducing soil compaction through the minimization of machinery footprint on arable land.

5.3.4 TILLAGE TIMING

Even the simple aspect of the timing of the tillage in a field can play a major factor in soil compaction. Early-season tillage is often performed to reduce weed density late in the season. However, early-season tillage often is the wrong choice for both soil compaction and weed control during the growing season. Late-season tillage allows for more organic material to be added to the soil while actively and drastically reducing the number of weeds present in the crop's growth cycle. Early tillage during wet spring times increases the soil's tendency toward compaction. Heavy equipment and traffic through the fields amplify the destruction of the soil's internal structure. Decreased pore space and limited soil and water volume can result from wet soil tillage during the early parts of the crop production season [51].

The impact of tillage operations during non-optimal wet conditions is a common concern for farm managers, and research into the actual implications of these kinds of operations is rather common. Figures 2.31 and 2.32 detail the results of a study looking into the change in resistance to soil penetration following tillage during wet conditions and the progression of soil aggregate strength throughout the growing season for these soils [52]. Figure 34 shows that after non-optimal cultivation, penetration resistance increased slightly compared to a reference soil that was not tilled, but that this resistance was still significantly lower than heavily

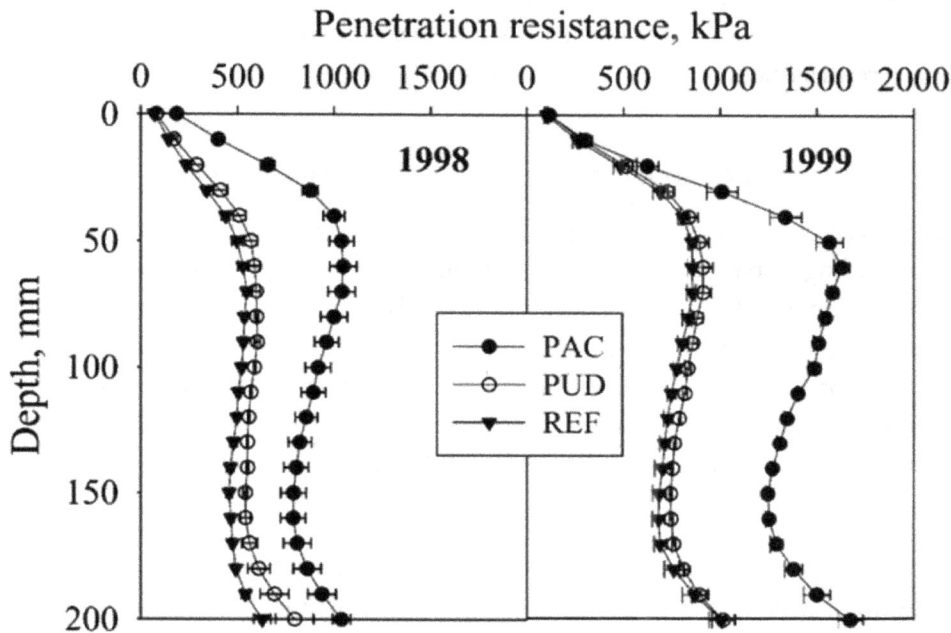

FIGURE 2.31. Soil penetration resistance measured shortly after tillage operations in May 1998 and 1999. PAC, compacted; PUD, intensive rotary cultivation [52].

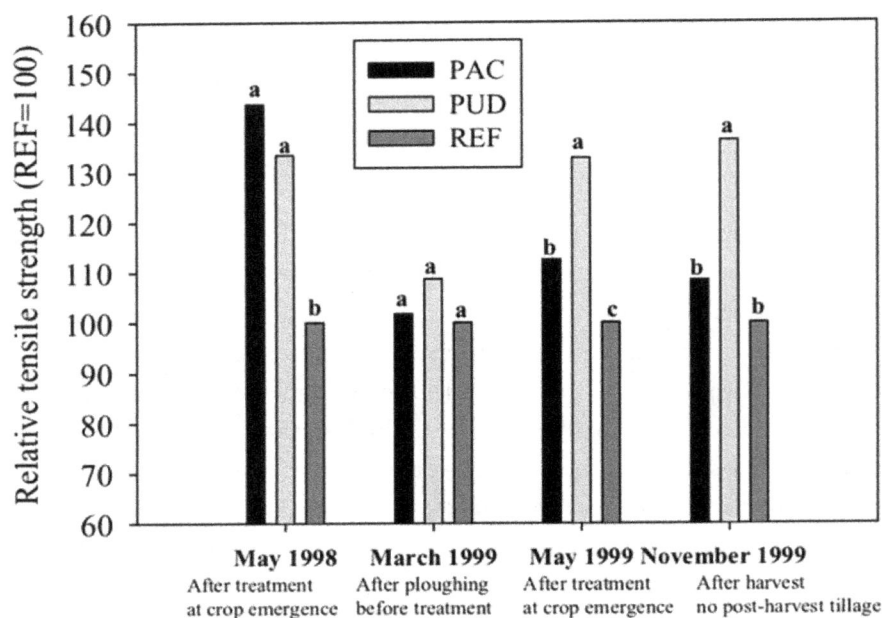

FIGURE 2.32. Relative tensile strength (REF = 100) of air-dried soil aggregates (average of the four size fractions) at the different times of sampling. PAC, compacted; PUD, intensive rotary cultivation [52].

compacted soil. The true consequences of non-optimal tillage operations are exposed in Figure 35, in which it is demonstrated that the tilled soil is unable to recover during the following growing cycle. As a result, the tilled soil maintains a very high soil aggregate tensile strength over time, further decreasing the soil's productivity and the long-term sustainability of agricultural operations in such soil.

5.4 COVER CROPPING AND CROP ROTATION

The final aspects of farm management that impact soil compaction and soil health are the decisions that farm managers make regarding crop rotation and cover cropping. Both have specific impacts for nutrient availability and storage, organic material availability and control, weed control, and erosion prevention. However, both cover cropping and crop rotation can also impact the prevention of soil compaction. This section will review the impacts of cover cropping versus crop rotation and discuss cover crop selection to achieve maximum compaction prevention and maintain the necessary levels of erosion prevention and the impact of preplanting cultivation and its effects on seedbed, germination, and root development.

5.4.1 COVER CROPPING VERSUS CROP ROTATION

Cover cropping is the practice of planting legume and grass varieties after the primary harvest, in the late fall, winter, or early spring before planting. Typically, these cover crops are planted to instill nutrients into the soil, increase the organic material in the topsoil layer, and better hold the soil together during tillage to prevent erosion issues. In addition to promoting yield advantages, cover cropping can also be used to

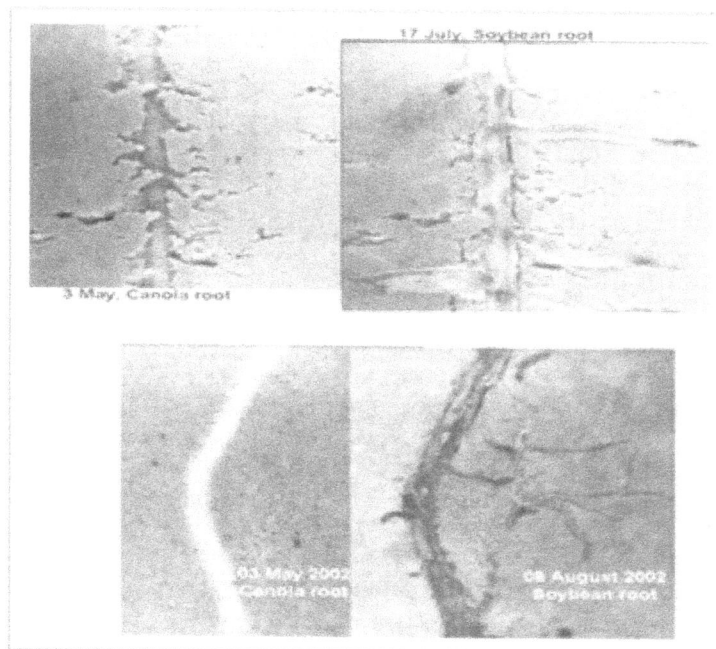

FIGURE 2.33. Minirhizotron images showing canola roots growing in May (left) and soybean roots observed in July and August (right) following the channels made by the preceding canola cover crop at 38.2 *cm* (at the WREC) (top) and 18 *cm* (at the Beltsville Agricultural Research Center) (bottom) depth. The bulk density was 1.55 and 1.61 *g/cm³*, and penetration resistance was 2247 and 2176 *kPa* for the upper and lower soils, respectively [54].

improve the soil profile and decrease existing compaction through the creation of pores and reduction of soil bulk density.

Crop rotation aims more at cycling specific nutrients within the soil matrix to promote a greater yield for specific crop types during different cyclic years. A good example of this is the common corn and soybean rotation, in which soybeans are rotated in when soil nutrient sampling indicates low nitrogen levels. Soybeans are utilized in this way due to their nitrogen-fixing attributes. This locks excess atmospheric nitrogen beyond what is needed for the soybean crop into the soil, to be used by corn in the following years. Crop rotation can additionally impact topsoil and subsurface soil compaction because of the differences in root penetration profiles. This can aid in moisture uptake and retention.

One study looked at the difference between cover cropping and crop rotation and then compared the impact on yield results as well as the resulting soil nutrients [53]. The findings were such that in the short term, there was little evidence to say that cover cropping alone could result in an adequate yield improvement, but a combination of cover cropping and crop rotation promoted increased crop yields and retained the benefits of using cover crops. On the other hand, when examining the effects on soil compaction, the long-term consequences of cover cropping helped to dramatically negate long-term compaction issues. Cover cropping plays an essential role in decreasing soil compaction through the reduction of soil bulk density, the alteration of soil aggregate size, the creation of root channels, and improving the aeration and pore space within the soil. Specific cover cropping can also help to combat long-term compaction by promoting subsoil disturbances via root channels.

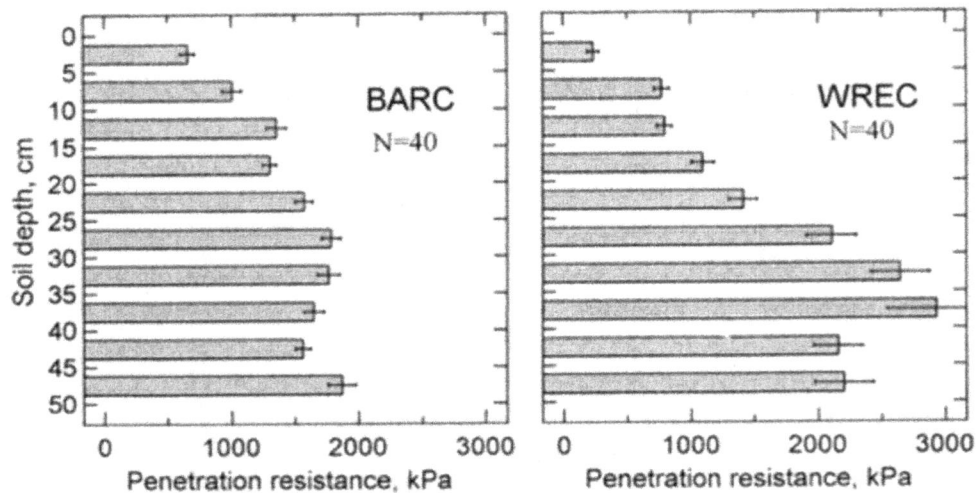

FIGURE 2.34. Penetration resistance (*kPa*) with depth (*cm*) at the Beltsville Agricultural Research Center and the WREC. The average volumetric water content at time of penetration resistance measurement was 0.22 *cm³/cm³* (WREC) and 0.27 *cm³/cm³* (Beltsville Agricultural Research Center) in the surface soil (0–20 *cm*) and 0.29 *cm³/cm³* (WREC) and 0.39 *cm³/cm³* (BARC) in the subsoil (20–40 *cm*) [55].

5.4.2 COVER CROP SELECTION

One of the primary ways in which cover crops can impact soil compaction is the creation of pore space and root channels. These openings help to decrease the soil's bulk density, break up previously compacted volumes, and promote water infiltration, all of which further aid in this endeavor. Figures 2.33 and 2.34 depict the results of studies on the effects of root profiles and root penetration resistance, which indicate compaction relief from cover cropping [54 and 55]. In particular, the studies investigated the differences in channels created by soybean and canola plant roots as well as the effects on soil nutrient and water content from a variety of other legume-type cover crops [54 and 55]. Cover cropping with radish and legume-type crops aided in decreasing soil penetration resistance during later planting and marginally disrupted soil compaction. In addition, cover cropping had added benefits for nutrient content and water availability. Data from the Wye Research and Education Center (WREC), in Figure 40, showed how cover cropping impacted soil with historically high compaction [55]. Utilizing cover crops with large root profiles was particularly effective at increasing the macroporosity and facilitating aggregate break-up in both topsoil and subsoil [54]. The latter is particularly important for soil types with increased risk of compaction, such as those with a high clay content. In addition to its other benefits, cover cropping is a useful and inexpensive method that aids in alleviating the effects of previous compaction, costing far less than mechanical relief applied through subsoiling operations.

6. CONCLUSION

This chapter provided a comprehensive review of the soil compaction problem. The historical aspects, the mechanism, and the environmental implications of soil compaction were discussed. A large off-road vehicle maneuvering with a draft load causes the soil beneath each point to compact until it can support the imposed load in both the vertical and horizontal directions. Different vehicle designs have advantages and disadvantages in addressing the soil compaction problem. The track-versus-tires debate continues to this day, and the farmer's choice should depend on their specific situation. Sophisticated farm management practices can significantly reduce soil compaction in the midterm. Farmers risk losing 50% of their expected profits when soil compaction is not addressed in a sustainable way. Farmers and policymakers are encouraged to work toward reducing and reversing soil compaction for sustainable management of agricultural lands.

The issue of soil compaction is not one that will ever cease to exist. It will continue to cause trouble for those in agriculture, construction, mining, and other industries that deal with soil and ground working. As such, it is important that an understanding of the impact of soil compaction is continually being disseminated into these industries as well as the basic management practices that can help to prevent an extensive spread of the problem. For design engineers in these fields, soil compaction offers the potential for continued improvement in equipment design. Looking specifically at agriculture, the ongoing trend of increasing equipment size and capacity in order to improve fuel and energy sustainability indicates that there will be a continued demand to further reduce the equipment loading and improve soil interaction of crop machinery in order to maintain an adequate level of soil compaction minimization. The design of tillage equipment has already come a long way from the moldboard plow, specifically in terms of minimizing soil disturbed unnecessarily while maximizing the implement's capacity to pulverize the soil aggregates within the seedbed. Moving forward, tillage equipment's most likely challenge will be ensuring adequate wheel support during operation without causing additional soil loading and compaction forces.

Farm managers must recognize that the prevention of unnecessary soil compaction is of paramount interest in the long-term productivity of their resources. They need to adopt a continuous improvement attitude and do whatever is feasible to minimize compaction. The seemingly small benefits of tillage cycling, crop rotation, and cover cropping should not be overlooked, since these practices continue to prevent soil compaction and reduce equipment traffic in the fields. As with many other aspects of off-road vehicle and machine design, committing to improving the performance of all factors will increase the effectiveness of the soil compaction control and prevention areas. Because the ability to sustainably grow food is critical to humanity's future, agricultural engineers of the 21st century with a working knowledge of soil compaction phenomena will continue to be in high demand.

7. ACKNOWLEDGMENTS

We would like to acknowledge our fellow classmates from the fall 2021 Design of Off-Road Vehicles class at Purdue University's School of Agricultural and Biological Engineering for their contributions to the structure and content of this technical chapter. Robert M. Stwalley IV is acknowledged for his work on the figures within the document.

8. CONFLICT OF INTEREST

The authors declare no conflict of interest.

9. SOIL COMPACTION QUESTIONS

1. Describe how field traffic affects soil health and reduces the long-term sustainability of crop production in earthen media.

2. What current agricultural practices can lead to severe soil compaction?

3. Describe the advantages and disadvantages of standard tires, tracks, and wide floatation tire support and traction systems for agricultural equipment. What additional considerations do nonagricultural machines have?

4. List five harmful agricultural practices that can increase soil compaction. List three modern practices that can minimize soil compaction.

5. Describe two remediation processes that can reduce soil compaction that has already been created.

10. REFERENCES

[1] M. Lane, *The Story of the Steam Plough Works: Fowlers of Leeds*, Northgate Publishing Co., 1980.

[2] W. A. Raney, T. W. Edminster, and W. H. Allaway, "Current Status of Research in Soil Compaction," *Soil Science Society of America Journal*, vol. 19, pp. 423–428, 1955, https://doi.org/10.2136/sssaj1955.03615995001900040008x.

[3] TractorData, "Challenger 65," 2016. [Online]. [Accessed November, 29, 2021], from https://www.tractordata.com/farm-tractors/000/9/1/917-challenger-65.html.

[4] C. Hadjilambrinos, "Reexamining the Automobile's Past: What Were the Critical Factors That Determined the Emergence of the Internal Combustion Engine as the Dominant Automotive Technology?," *Bulletin of Science, Technology, and Society*, vol. 41, no. 2–3, p. 58–71, 2021, https://doi.org/10.1177/02704676211036334.

[5] B. D. Soane and C. van Ouwerkerk, "Soil Compaction Problems in World Agriculture," in *Developments in Agricultural Engineering*, vol. 11, Elsevier B.V., 1994, p. 1–21, https://doi.org/10.1016/B978-0-444-88286-8.50009-X.

[6] C. C. Spence, *God Speed the Plow: the Coming of Steam Cultivation to Great Britain*, University of Illinois Press, 1960.

[7] R. Horn, H. Domzzal, A. Słowińska-Jurkiewicz, and C. van Ouwerkerk, "Soil Compaction Processes and Their Effects on the Structure of Arable Soils and the Environment," *Soil and Tillage Research*, vol. 35, pp. 23–36, 1995, https://doi.org/10.1016/0167–1987(95)00479-C.

[8] S. Duiker, "Avoiding Soil Compaction," 2005. [Online]. Available: https://extension.psu.edu/effects-of-soil-compaction [Accessed November 21, 2021].

[9] T. Batey, "Soil Compaction and Soil Management—a Review," *Soil Use and Management*, vol. 25, no. 4, pp. 335–345, 2009, https://doi.org/10.1111/j.1475–2743.2009.00236.x.

[10] G. F. Botta, A. Tolon-Becerra, F. Bienvenido, D. Rivero, D. A. Laureda, A. Ezquerra-Canalejo, and E. E. Contessotto, "Sunflower Harvest: Tractor and Grain Chaser Traffic Effects on Soil Compaction and Crop Yields," *Land Degradation & Development*, vol. 29, pp. 4252–4261, 2018, https://doi.org/10.1002/ldr.3181.

[11] B. D. Soane and C. van Ouwerkerk, "Implications of Soil Compaction in Crop Production for the Quality of the Environment," *Soil and Tillage Research*, vol. 35, pp. 5–22, 1995, https://doi.org/10.1016/0167–1987(95)00475–8.

[12] H. A. Torbert and C. W. Wood, "Effects of Soil Compaction and Water-Filled Pore Space on Soil Microbial Activity and N Losses," *Communications in Soil Science and Plant Analysis*, vol. 23, p. 1321–1331, 1992, https://doi.org/10.1080/00103629209368668.

[13] G. F. Botta, A. Tolon-Becerra, M. Tourn, X. Lastra-Bravo, and D. Rivero, "Agricultural Traffic: Motion Resistance and Soil Compaction in Relation to Tractor Design and Different Soil Conditions," *Soil and Tillage Research*, vol. 120, pp. 92–98, 2012, https://doi.org/10.1016/j.still.2011.11.008.

[14] M. R. Shaheb, R. Venkatesh, and S. A. Shearer, "A Review of the Effect of Soil Compaction and Its Management for Sustainable Crop Production," *Journal of Biosystems Engineering*, vol. 46, pp. 417–439, 2021, https://doi.org/10.1007/s42853-021-00117-7.

[15] W. B. Voorhees, "The Effect of Soil Compaction on Crop Yield," *SAE Transactions*, vol. 95, no. 3, pp. 1078–1084, 1986, url: https://www.jstor.org/stable/44725467.

[16] H. Jamail, G. Nachimuthy, B. Palmer, D. Hodgson, A. Hundt, C. Nunn, and M. Braunack, "Soil Compaction in a New Light: Knowing the Cost of Doing Nothing—A Cotton Case Study," *Soil & Tillage Research*, vol. 213, p. 105158, 2021.

[17] T. Chamen, "Controlled Traffic Farming—From Worldwide Research To adoption in Europe and Its Future Prospects," *Acta Technologica Agrriculturae*, vol. 18, no. 3, pp. 64–73, 2015, https://doi.org/10.1515/ata-2015-0014.

[18] A. Zabrodskyi, E. Sarauski, S. Kukharets, A. Juostas, G. Vasiliauskas, and A. Andriusis, "Analysis of the Impact of Soil Compaction on the Environment and Agricultural Economic Losses in Lithuania and Ukraine," *Sustainability*, vol. 13, p. 7762, 2021, https://doi.org/10.3390/su13147762.

[19] K. Gunjal, G. Lavoie, and G. S. V. Raghavan, "Economics of Soil Compaction due to Machinery Traffic and Implications for Machinery Selection," *Canadian Journal of Agricultural Economics/Revue canadienne d'agroeconomie*, vol. 35, pp. 591–603, 1987, https://doi.org/10.1111/j.1744-7976.1987.tb02251.x.

[20] A. N. Shah, M. Tanveer, B. Shahzad, G. Yang, S. Fahad, S. Ali, M. A. Bukhari, S. A. Tung, A. Hafeez, and B. Souliyanonh, "Soil Compaction Effects on Soil Health and Crop Productivity: An Overview," *Environmental Science and Pollution Research*, vol. 24, pp. 10056–10067, 2017, https://doi.org/10.1007/s11356-017-8421-y.

[21] U. Ghosh and A. L. Daigh, "Soil Compaction Problems and Subsoiling Effects on Potato Crops: A Review," *Crop, Forage and Turfgrass Management*, vol. 6, pp. 1–10, 2020, https://doi.org/10.1002/cft2.20030.

[22] P. G. Mellgren, "Terrain Classification for Canadian Forest," *Canadian Pulp and Paper Association Journal*, pp. 1–13, 1980.

[23] M. F. Nawaz, G. Bourrie, and F. Trolard, "Soil Compaction Impact and Modelling: A Review," *Agronomy for Sustainable Development*, vol. 33, p. 291–309, 2013, https://doi.org/10.1007/s13593-011-0071-8.

[24] C. W. Smith, M. A. Johnston, and S. Lorentz, "Assessing the Compaction Susceptibility of South African Forestry Soils: I. The Effect of Soil Type, Water Content and Applied Pressure on Uni-axial Compaction," *Soil and Tillage Research*, vol. 41, pp. 53–73, 1997, https://doi.org/10.1016/S0167-1987(96)01084-7.

[25] P. Jasa, "Avoiding Harvest Compaction in Wet Soils," University of Nebraska-Lincoln Institute of Agriculture and Natural Resources, 2019. [Online]. Available: https://cropwatch.unl.edu/2019/avoiding-compaction-harvest. [Accessed November 29, 2021].

[26] C. Shawver, J. Brummer, J. Ippolito, J. Ahola, and R. Rhoades, "Managing Cattle Impacts When Grazing on Wet Soils," 2020. [Online]. Available: https://extension.colostate.edu/topic-areas/agriculture/managing-cattle-impacts-when-grazing-on-wet-soils-1-634/. [Accessed November 28, 2021].

[27] National Inventors Hall of Fame, "Benjamin Holt Track-Type Tractor," 2006. [Online]. Available: https://www.invent.org/inductees/benjamin-holt. [Accessed November 30, 2021].

[28] Caterpillar, "1900s," Caterpillar, Inc., 2021. [Online]. Available: https://www.caterpillar.com/en/company/history/1900.html. [Accessed November 26, 2021].

[29] Case IH, "Steiger & Quadtrac," CNH Industrial America, LLC, 2021. [Online]. Available: https://www.caseih.com/emea/en-za/products/tractors/steiger-quadtrac-series/steiger-quadtrac. [Accessed November 23, 2021].

[30] E. M. Hawkins, "Benchmarking Costs of Fixed-Frame, Articulated, and Tracked Tractors," *Applied Engineering in Agriculture*, vol. 31, no. 5, pp. 741–745, 2015, https://doi.org/10.13031/aea.31.11074.

[31] M. Lamandé, M. H. Greve, and P. Schjønning, "Risk Assessment of Soil Compaction in Europe—Rubber Tracks or Wheels on Machinery," *Catena*, vol. 167, pp. 353–362, 2018, https://doi.org/10.1016/j.catena.2018.05.015.

[32] Elmer's Manufacturing, Inc., "The Benefits of Tracks vs. Tires," 2016. [Online]. Available: https://elmersmfg.com/2016/04/benefits-tracks-vs-tires/. [Accessed November 24, 2021].

[33] NTS Tire Supply Team, "Tires vs. Tracks: Which Creates Less Compaction?," 2019. [Online]. Available: https://www.ntstiresupply.com/ptk-shared/tires-vs-tracks-which-creates-less-compaction. [Accessed December 1, 2021].

[34] P. Norman, "IF and VF Tyres—What Are They?," Brocks Wheel & Tyre, 2021. [Online]. Available: https://bwt.uk.com/news/if-tyres-and-vf-tyres-what-are-they/. [Accessed November 21, 2021].

[35] J. Tuschner, "Compare & Contrast—Making the Case for Tires vs. Tracks," Farm Equipment, 2020. [Online]. Available: https://www.farm-equipment.com/articles/18193. [Accessed November 30, 2021].

[36] T. Keller and J. Arvidsson, "Technical Solutions to Reduce the Risk of Subsoil Compaction: Effects of Dual Wheels, Tandem Wheels and Tyre Inflation Pressure on Stress Propagation in Soil," *Soil and Tillage Research*, vol. 79, pp. 191–205, 2004, https://doi.org/10.1016/j.still.2004.07.008.

[37] Titan International, Inc., "LSW Technology," 2014. [Online]. Available: https://www.titan-intl.com/innovation/lsw-tires. [Accessed December 2, 2021].

[38] M. Lyasko, "Slip Sinkage Effect in Soil-Vehicle Mechanics," *Journal of Terramechanics*, vol. 47, pp. 21–31, 2010, https://doi.org/10.1016/j.jterra.2009 .08.005.

[39] N. H. Abu-Hamdeh, T. G. Carpenter, R. K. Wood, and R. G. Holmes, "Soil Compaction of Four-Wheel Drive and Tracked Tractors under Various Draft Loads," in *International Off-Highway & Powerplant Congress & Exposition*, Warrendale, PA, 1995, https://doi.org/10.4271/952098.

[40] A. Moinfar, G. Shahgholi, Y. Abbaspour-Gilandeh, I. Herrera-Miranda, J. L. Hernandez-Hernandez, and M. A. Herrera-Miranda, "Investigating the Effect of the Tractor Drive System Type on Soil Behavior under Tractor Tires," *Agronomy*, vol. 11, no. 4, 2021, https://doi.org/10.3390/agronomy11040696.

[41] D. B. Davies, J. B. Finney, and S. J. Richardson, "Relative Effects of Tractor Weight and Wheel-Slip in Causing Soil Compaction," *Journal of Soil Science*, vol. 24, pp. 399–409, 1973, https://doi.org/10.1111/j.1365–2389.1973. tb00775.x.

[42] V. V. Vantsevich and M. V. Blundell, eds., *Advanced Autonomous Vehicle Design for Severe Environments*, IOS Press, 2015.

[43] S. Kassel, "Lunokhod-1 Soviet Lunar Surface," RAND Corporation, Santa Monica, 1971.

[44] "Perseverance (Rover)," Wikipedia, 2021. [Online]. Available: https://en.wikipedia.org/wiki/Perseverance_(rover). [Accessed November 26, 2021].

[45] S. H. Hoseinian, A. Hemmat, A. Esehaghbeygi, G. Shahgoli, and A. Baghbanan, "Development of a Dual Sideway-Share Subsurface Tillage Implement: Part 2. Effect of Tool Geometry on Tillage Forces and Soil Disturbance Characteristics," *Soil and Tillage Research*, vol. 215, p. 105200, 2021, https://doi.org/10.1016/j.still.2021.105200.

[46] G. Spoor, "Alleviation of Soil Compaction: Requirements, Equipment and Techniques," *Soil Use and Management*, vol. 22, pp. 113–122, 2006, https://doi.org/10.1111/j.1475–2743.2006.00015.x.

[47] V. I. Adamchuk, A. V. Skotnikov, J. D. Speichinger, and M. F. Kocher, "Development of an Instrumented Deep-Tillage Implement for Sensing of Soil Mechanical Resistance," *Transactions of the ASAE*, vol. 47, no. 6, pp. 1913–1919, 2004, https://doi.org/10.13031/2013.17798.

[48] S. Winsor, "Healthy Soil and Profits from Low-Till," Farm Progress, 2012. [Online]. Available: https://www.farmprogress.com/tillage/healthy-soil-and-profits-low-till. [Accessed November 23, 2021].

[49] D. H. Alban, G. E. Host, J. D. Elioff, and D. A. Shadis, "Soil and Vegetation Response to Soil Compaction and Forest Floor Removal after Aspen Harvesting," USDA Forest Service, 1994, https://doi.org/10.2737/NC-RP-315.

[50] Misser Uitgeverij B.V., "NEXAT Redesigning Versatility, Reducing Field Traffic," 2021. [Online]. Available: https://www.futurefarming.com/tech-in-focus/autonomous-semiauto-steering/autonomous-vehicles/nexat-redesigning-versatility-reducing-field-traffic/. [Accessed 6 January 2022].

[51] R. L. Raper, D. W. Reeves, C. H. Burmester, and E. B. Schwab, "Tillage Depth, Tillage Timing, and Cover Crop Effects on Cotton Yield, Soil Strength, and Tillage Energy Requirements," *Applied Engineering in Agriculture*, vol. 16, no. 4, pp. 379–385, 2000, https://doi.org/10.13031/2013.5363.

[52] L. J. Munkholm and P. Schjonning, "Structural Vulnerability of a Sandy Loam Exposed to Intensive Tillage and Traffic in Wet Conditions," *Soil and Tillage Research*, vol. 79, pp. 79–85, 2004, https://doi.org/10.1016/j.still.2004.03.012.

[53] I. A. Dozier, G. D. Behnke, A. S. Davis, E. D. Nafziger, and M. B. Villamil, "Tillage and Cover Cropping Effects on Soil Properties and Crop Production in Illinois," *Agronomy Journal*, vol. 109, pp. 1261–1270, 2017, https://doi.org/10.2134/agronj2016.10.0613.

[54] J. C. Calonego, J. P. Raphael, J. P. Rigon, L. de Olieria-Neto, and C. A. Rosolem, "Soil Compaction Management and Soybean Yields with Cover Crops under No-till and Occasional Chiseling," *European Journal of Agronomy*, vol. 85, pp. 31–37, 2017, https://doi.org/10.1016/j.eja.2017.02.001.

[55] S. M. Williams and R. R. Weil, "Crop Cover Root Channels May Alleviate Soil Compaction Effects on Soybean Crop," *Soil Science Society of America Journal*, vol. 68, p. 1403–1409, 2004, https://doi.org/10.2136/sssaj2004.1403.

CHAPTER 3. COMPOSITE BEARINGS

All categories of off-road vehicles typically operate in less than pristine environments, which are loaded with all types of potential contaminants. Basic metal components can be adequately protected with a coating of heavy-duty paint, but bearing assemblies are more delicate. They contain some form of lubricant and provide a motion interface for parts. Contamination is detrimental to their overall life, and many bearings are in difficult-to-access places on off-road equipment. If sealed, they will eventually need replacement when the lubricant runs dry. If zerked, they require periodic grease additions to make up for the consumption from movement. Both alternatives can be problematic if the bearing is buried deep within the machine. Composite bearings represent a modern technology that does not require periodic greasing or replacement. These assemblies carry their lubricants in the base materials used to create the bearing assembly and have expected lifetimes exceeding normal sealed bearings. This overview of composite bearings was designed to acquaint equipment designers with the advantages of this new technology for use in the heavy-duty off-road machinery realm. This manuscript was originally accepted and intended for publication in the book *Tribology and Predictive Maintenance—Recent Advances*, but it was transferred to *Composite Materials: Science and Engineering* when the original book was undersubscribed on contributions.

THE UTILIZATION OF COMPOSITE BEARINGS IN HEAVY AGRICULTURAL, CONSTRUCTION, FORESTRY, AND MINING EQUIPMENT DESIGN APPLICATIONS

THOMAS BARNES, PURDUE UNIVERSITY AGRICULTURAL AND BIOLOGICAL ENGINEERING

BRECKEN BEYER, PURDUE UNIVERSITY AGRICULTURAL AND BIOLOGICAL ENGINEERING

WYATT GRIFFEY, PURDUE UNIVERSITY MECHANICAL ENGINEERING

CALEB HUFFMEYER, PURDUE UNIVERSITY AGRICULTURAL AND BIOLOGICAL ENGINEERING

CLAYTON NESS, PURDUE UNIVERSITY AGRICULTURAL AND BIOLOGICAL ENGINEERING

OWEN NIFONG, PURDUE UNIVERSITY AGRICULTURAL AND BIOLOGICAL ENGINEERING

TYLER J. MCPHERON, PURDUE UNIVERSITY AGRICULTURAL AND BIOLOGICAL ENGINEERING

ROBERT M. STWALLEY III, PURDUE UNIVERSITY AGRICULTURAL AND BIOLOGICAL ENGINEERING

ABSTRACT

This design overview will describe the fundamental characteristics of composite material bearings and how they should be utilized in the design of agricultural, construction, forestry, and mining heavy equipment. Currently, metallic bearings are used in a multitude of heavy machinery and implement applications, where the movement of mechanical parts with reduced friction in a single-degree of articulated motion is required. Most metallic bearings use grease or oil for lubrication to create a near-frictionless motion. Off-road heavy machinery uses many different types of metallic bearings in their designs to achieve their specialized functions. However, metallic bearings generally require periodic maintenance to sustain their lubricity after repetitive use, and they are prone to corrosion. Maintenance intervals are frequent and increase operational downtime for equipment operators and technicians, which further increases operating costs and reduces productivity. Composite bearings are manufactured from fiber-reinforced materials and are produced using a variety of resin formulas to achieve the desired composite cross-section. This present work focuses on potential design applications of self-lubricated composite bearings using polytetrafluoroethylene or other embedded lubrication materials to minimize friction coefficients. Composite bearings are advantageous in heavy equipment applications due to their greaseless design, corrosion resistance, high bearing load capacities, and resistance to fatigue loads and impact. Further secondary advantages of composite bearings over traditional metallic bearings are described. Durability and design life tests completed by major composite bearing manufacturers, in which minimal wear was measured after one million to two million cycles, were referenced to illustrate the longevity of composite bearing materials. Equipment in the agricultural, construction, forest, and mining industries operates in harsh outdoor environments, and they are designed with heavy-duty components that resist or tolerate these conditions. A further understanding of composite bearing characteristics will allow engineers to more frequently incorporate them into their designs to reduce maintenance costs, improve machine productivity, and accommodate adverse environmental conditions, all while minimizing bearing wear and friction created by mechanical motion.

KEYWORDS: composite bearings, durability, friction, design engineering, agriculture, construction, forestry, mining, equipment manufacturing

1. INTRODUCTION

Bearings are utilized in a variety of mechanical designs and machinery applications to minimize friction coefficients and produce smooth fluid-like mechanical motion between two working components having a single degree of articulated or rotary motion. Although bearings are commonly used in many different industries, the agricultural, construction, forestry, and mining equipment industries incorporate an abundance of heavy-duty bearings in their vehicle designs. Compared to on-road vehicles, which have bearings in suspension components, engines, transmissions, drivetrains, and steering components, heavy off-road equipment typically features auxiliary motion functions that are unique to the specific piece of equipment, in addition to the same general parts that compose an on-road vehicle. For example, an agricultural combine harvester has many mechanical functions within its grain threshing system, and construction backhoe loaders have mechanical and hydraulic front loader buckets and backhoes, all of which require an array of bearings for frictionless, mechanical motion, and smooth operation.

Many different types of bearings will be examined in this overview, but there are two general categories of traditional metallic bearings incorporated in most mobile equipment: plain bearings and antifriction bearings. Plain bearings are those that have a sliding or linear movement between the bearing surface and the contact part. Antifriction bearings have an internal idler gear to facilitate rotational movement, such as a rotating power shaft and the yaw articulation of a backhoe boom [1]. Many bearings types feature mechanical designs to reduce friction coefficients, such as roller bearings, which use small cylinders inside of a steel cage to allow low-friction rotation. Almost all traditional metallic bearings use a form of lubricant as their main friction-minimizing substance. This is typically a high-viscosity grease or a lower-viscosity film of oil between the bearing surface and the contact material. Figure 3.1 shows an example of a plain journal bearing used for shaft rotation [2], while Figure 3.2 shows the differences between a roller antifriction bearing and a ball antifriction bearing [3].

Most bearings that are currently used in heavy equipment designs are made of metallic materials and require regular maintenance to sustain the performance and integrity of the bearing. This introduces a variety of issues for design engineers, equipment operators, and service technicians. Daily service requirements are detrimental to equipment productivity and operating costs due to the cost of grease and equipment downtime. Maintenance neglect can result in catastrophic failures, fires, and expensive repair costs. To mitigate this issue, equipment longevity research indicates that composite material bearings could be a potential solution. These bearings require minimal long-term maintenance compared with traditional metallic bearings in heavy equipment applications.

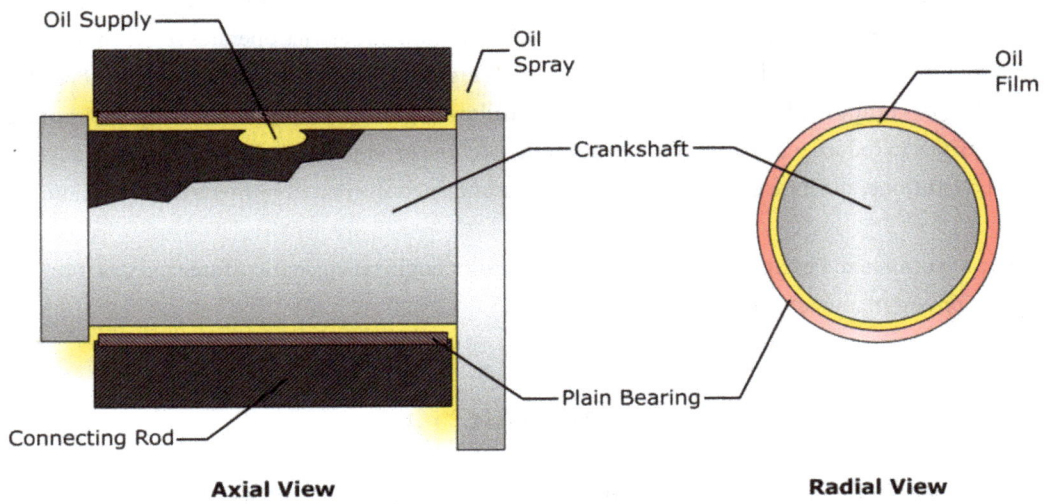

Oil Supply

Oil Spray

Crankshaft

Oil Film

Plain Bearing

Connecting Rod

Axial View

Radial View

FIGURE 3.1. Plain bearing diagram from side and in section [2].

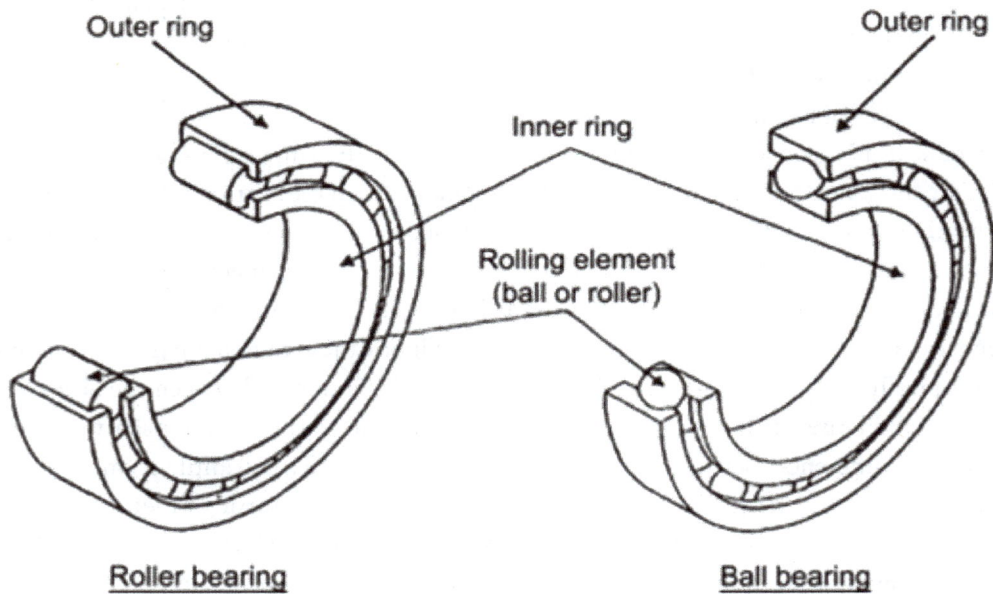

Outer ring

Outer ring

Inner ring

Rolling element (ball or roller)

<u>Roller bearing</u>

<u>Ball bearing</u>

FIGURE 3.2. Cross-section of antifriction bearings with rollers (left) and balls (right) [3].

Composite bearings utilize self-lubricating mechanisms that are embedded in the surface layer of the bearing's material during the manufacturing process. The embedded lubrication is released from the surface layer, which is typically a perforated or porous layer, when the shaft surface moves relative to the bearing surface layer. The creation of friction at the motion points causes the embedded oil to be released at the contact interface and provide a layer of lubrication between the material surface and the bearing surface. This is highly beneficial, because constant lubrication can be achieved independent of the speed of the rotating or linearly moving part [4]. Any grease or lubricant not used is reabsorbed back into the porous bearing surface when the component is at rest. Additionally, self-lubricating bearings do not lose or deteriorate their grease over time, and they do not require regular grease or oil service intervals to maintain their functional performance. This maximizes equipment uptime and productivity and minimizes maintenance costs.

2. LITERATURE REVIEW

Peer-reviewed journal publications discussing design recommendations and technical information for the application of composite bearings are available, but they seem limited [5–7]. Journals tend to primarily concentrate on composite bearing material properties and the development of composite materials [8–13]. However, many composite bearing manufacturers provide a wealth of technical information for their products, and they even include detailed design guides on their websites [14–17]. The specifications and critical details from these component design guides were heavily used to summarize and provide design recommendations in this work as well as more thorough descriptions of the general characteristics of composite bearings. Corporate-provided information from leading composite bearing manufacturers, such as Igus Motion Plastics, GGB by Timken, and Polygon Composites Technology, was used to describe different mechanical properties, thermal properties, and the general bearing geometries of typical composite bearing assemblies.

Polygon Composites Technology has been manufacturing composite material products for nearly 70 years. Polygon produces composite tubing and self-lubricating composite bearings as well as other custom composite products for a variety of industries, such as agriculture, construction, medical, oil and gas, and on-road trucks. Polygon has documented several successful applications of composite bearings, whereas other bearing types were inferior or where the only solution was a custom-designed bearing [18]. Polygon describes design applications in the agricultural and construction sectors that were summarized for this heavy equipment–specific research, providing real-world examples of composite bearings replacing steel bearings, which optimized the overall design and decreased projected maintenance time. Another manufacturer, Igus Motion Plastics, manufactures chain and cable carriers, composite bearings, gears, and robotic equipment parts and validates its composite products in specific applications to ensure product quality and meet user quality specifications [19]. The manufacturers have dedicated labs for testing composite bearing products under a variety of different wear conditions and structural loads. Igus makes some of its testing data available to engineers and consumers to accurately compare products to determine the most ideal product for the intended application. Finally, GGB by Timken began making composite material bearings in the late 1970s. GGB produces a hybrid-style sleeve bearing that utilizes self-lubricating, composite technology on the inner layer, with a reinforced exterior steel layer for heavy-duty applications. The GGB composite bearing assemblies have been designed specifically to support vehicular applications [20].

Bearings design metrics covering all primary bearing types feature wear estimation calculations, and bearing load calculations for a variety of mechanical applications remain valid for composite bearings [1]. A thorough understanding of bearing lubrication, bearing failures, and bearing size selection is assumed to be within the reader's background. The current overview has utilized product literature, composite material testing, and design guides produced by composite bearing manufacturers to provide a practical overview and design guide for engineers to apply composite bearing technology in the heavy-duty agricultural, construction, forestry, and mining equipment industries. Since the composite bearing market is a relatively small part of the overall bearing industry, this overview focuses on leading manufacturers in the market. The technological advancements of their products are summarized and presented through their published technical literature.

3. TRADITIONAL METALLIC BEARINGS

There are two general types of bearings that provide axial support for a rotating shaft: plain bearings and antifriction bearings. In traditional designs, the selection of the bearing type would typically depend on the categorization of the motion, the type and direction of the loads, the application, and the environment. Regardless of application and type, traditional bearings can be selected to fit most reasonable design constraints. A brief analysis of traditional metallic bearings and their associated shortcomings for off-road vehicle and heavy equipment applications is presented in this section.

3.1 PLAIN BEARINGS

Traditional plain bearings, also known as journal bearings, such as the one shown in Figure 3.1, are made of many metallic materials but typically bronze or brass. There are a wide variety of plain bearings, but all consist of a lubricated metal sleeve designed to reduce friction between two sliding or rotating surfaces. This design, when operated properly, results in complete hydrodynamic contact between the shaft and bearing surfaces. These bearings are capable of a high load capacity and can operate at low speeds. Due to having fewer moving parts, these bearings are cheaper to manufacture and have low operating noise. Their simplistic design allows for a variety of bearing housing sizes and relative ease in maintenance or replacement.

The main disadvantages of the plain bearing design are its high-speed limitations and its continual high lubrication requirement, generally requiring a dedicated external lubrication system. The high-contact area improves the bearing's ability to sustain high loads but also decreases the ability to operate at high speeds, limiting potential applications. Failure to apply the proper amount of lubrication can increase friction between the moving parts, and this can lead to significant wear, especially during start-up [2]. Finally, the lubrication and maintenance requirements of a plain journal bearing system are major disadvantages due to the cost in time, additional system overhead, and machine service downtimes. While traditional metallic plain bearings have served the off-road industry for years, the disadvantages associated with them have also caused engineers to continue to look for improvements on this traditional bearing design.

3.2 ANTIFRICTION BEARINGS

To fight frictional loss and increase the wear life of bearings in adverse environments, antifriction bearings make use of small rollers inside of a steel cage to act as idlers between the rotating and fixed surfaces. The basic components of a roller and a ball bearing were shown in Figure 3.2. Antifriction bearings perform well in high-load, high-speed environments. Other advantages of this design include accurate shaft alignment, low starting friction, reduced rolling torque, and a simple process for lubrication and bearing replacement [1].

3.2.1 BALL BEARINGS

Ball bearings are the main form of antifriction bearings. Ball bearings utilize spherical rollers to decrease the friction between rotating inner races and stationary outer races. The outer race is generally held stationary by a bearing holder, and the inner race holds the shaft or pin. This design allows for ball bearings to operate at higher speeds than traditional plain bearings. Ball bearings can operate under radial and thrust loads, but they cannot provide significant axial load unless used in a tapered configuration, shown in Figure 3.3. The addition of the conical bearing area to the outer race provides axial load support into the cone. The doubling or tripling of ball rows, as shown in Figure 3.4, can also increase the load support capability.

The main disadvantages of ball and roller bearings are the lubrication requirement and reduced load-carrying capacity in relation to plain bearings. The reduced load capacity comes from the smaller contact area between the balls or rollers and the races in the bearings, compared to plain bearings, which carry load across their entire engagement interface. Lubrication concerns are a major limitation for ball bearings in off-road equipment. These limitations are part of the design of these bearings and cannot be overcome.

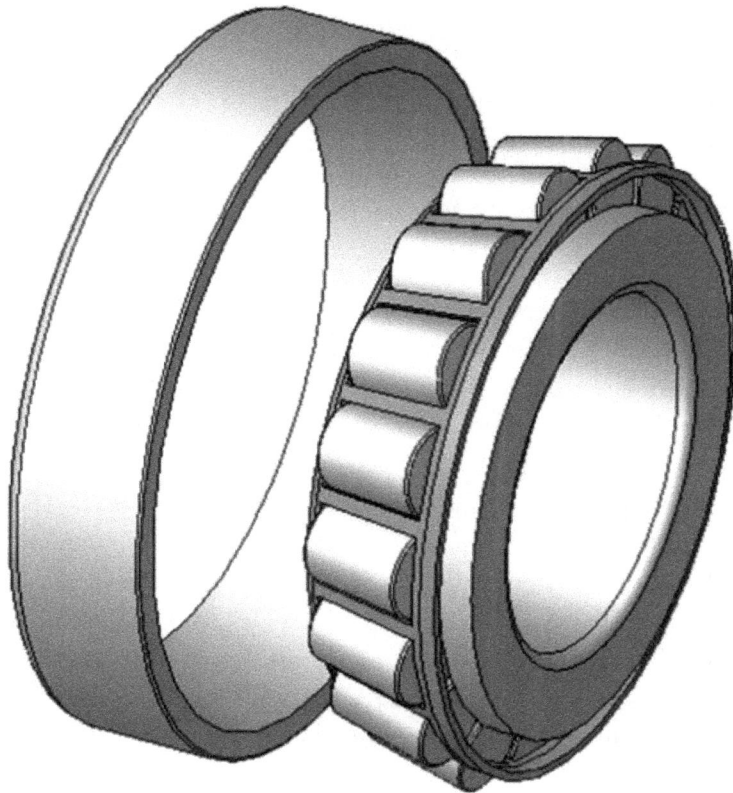

FIGURE 3.3. Exploded view of tapered roller bearing assembly, showing the conical section to carry axial load.

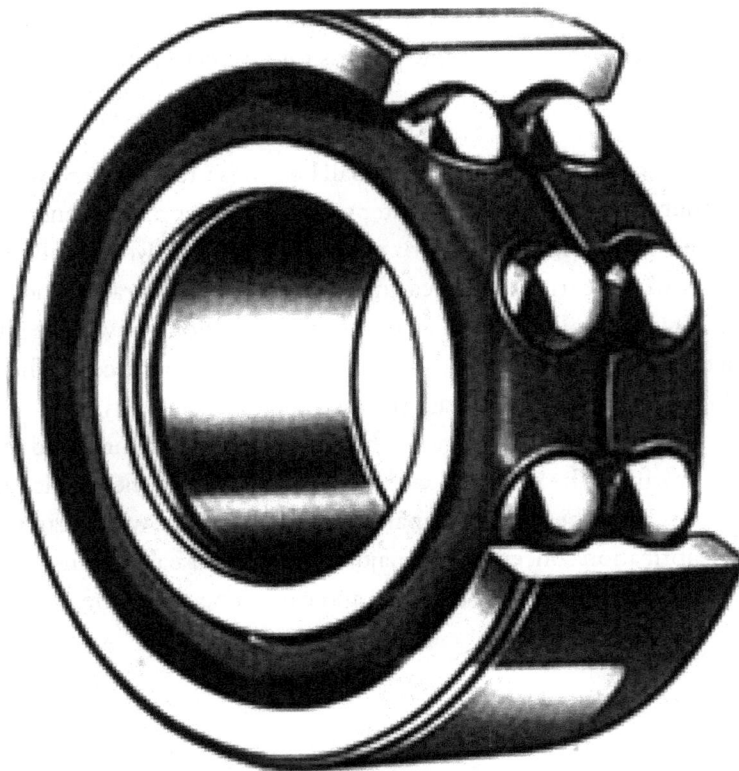

FIGURE 3.4. Cutaway view of a double row of ball rollers in a bearing assembly.

3.2.2 ROLLER BEARINGS

Roller bearings provide several advantages over both plain bearings and ball bearings, particularly in axial loading applications. The cylindrical rollers allow for more contact area than balls provide, which means that a larger load can be supported. Taper allows the bearing to support a load in the radial and axial directions, and this makes tapered roller bearings a common choice for wheels in the hubs and tracks on heavy equipment. The disadvantages of roller bearings include abnormal stresses resulting from the tapered design and the precision machining necessary to manufacture them. Adequate lubrication for roller bearing performance is vital, since improper greasing risks seal failure and ultimately causes premature bearing failure. Although the required maintenance is relatively simple, it does create machine downtime, additional labor costs, and lubricant expenses. These maintenance needs can represent significant losses in the field capacity for off-road vehicles. Productive field capacity is lost when the equipment is not in operation due to maintenance requirements. Improper greasing by applying too little grease can lead to significant metal-to-metal contact, grinding between components, and overheating, resulting in permanent damage of the bearing. Overgreasing of a bearing can led to damage in the bearing seals, creating excess drag through fluid friction, and other potential environmental concerns due to excessive lubricant being ejected from the bearing [21]. Finally, an ineffective seal will lead to bearing contamination from solid particles or water, and this circumstance significantly reduces bearing life and will eventually cause bearing failure. Plain and antifriction bearings are useful, but their shortcomings have caused engineers to look for methods to improve upon the limitations of traditional metallic bearings.

4. COMPOSITE BEARINGS

In general, a composite bearing looks a great deal like a plain journal bearing, with perhaps a smaller radial clearance with the shaft. Instead of needing an external lubrication source, composite bearings have an internal source, based on the bearing material containing a lubricant within its material matrix and the base material being slippery itself. This section will examine useful composite materials, lubricants, load-carrying ranges, bearing thermal properties, and manufacturing tolerances.

4.1 MATERIALS

Composite bearings resemble journal bearings. Like plain bearings, composite bearings utilize a lubrication film to minimize friction. However, unlike traditional bearings, composite bearings do not require an external lubricant source to provide the slippery film that coats the contacting surfaces. Composite bearings often utilize self-lubricating mechanisms that are embedded in the surface layer of the bearing during the manufacturing process. The surface layer is the main sliding contact interface between the two independent parts. The embedded lubrication is released from the sliding layer, which is typically a perforated or porous layer, when the material surface moves against the bearing surface layer. The creation of friction causes the embedded oil to coat the surface contact material and provide a layer of lubrication between the material surface and the bearing surface. Some common self-lubricating materials found in composite bearings are provided in Table 3.1 [22].

The sliding contact layer is typically composed of a tightly wound polytetrafluoroethylene (PTFE) plastic and a high-strength fiber layer that is encapsulated in a self-lubricating, high-temperature epoxy resin. The type of PTFE, the fiber material, and the lubrication-filled epoxy resin utilized vary by application and bearing manufacturer. To provide additional physical support to the bearing, a stiff backing layer will typically be meshed with the inside sliding contact layer. This backing layer is generally composed of continuously wound fiberglass coated with a high-temperature, high-stress epoxy. The ratio between sliding and backing layers that comprises the bearing varies by application and the composition of the sliding layer. A typical example of the surface and sliding layers is shown in Figure 3.5 [23].

TABLE 3.1. *Common materials used in manufacturing composite bearings (Carrera, 2023) [22]*

Rulon	Delrin AF
Fluorosint	TriSteel PT/PI
PTFE Blends	Ultracomp
Graphite PI	Nylon 6/6, 6/12
Ertalyte Tx	UHMW

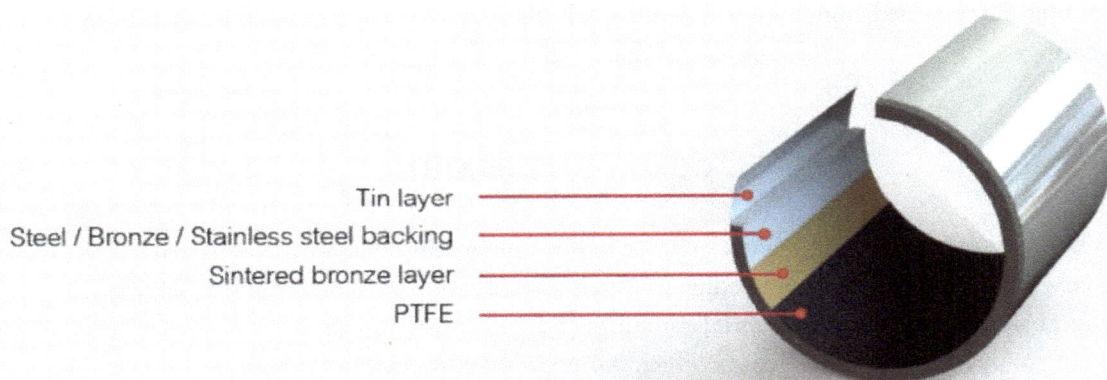

FIGURE 3.5. Antifriction bearing showing sliding layer, surface layer, and backing layers [23].

4.2 LUBRICATION AND FRICTION

Most composite bearings do not need lubrication. PTFE bearings fall into this category. This is due to the inherent dry sliding characteristics of the materials used in manufacturing composite bearings. Some component assemblies do require an initial charge of lubrication at installation. Polyoxymethylene (POM) composite bearings are designed to have an initial film of lubrication and do not require any further application for the life of the bearing. The PTFE and POM composite bearings have similar friction coefficients, which vary between 0.03 and 0.25 depending on external factors and operating conditions. It is important to note that POM composite bearings tend to have lower friction coefficients due to their initial application of lubricant. Figure 3.6 shows how operating conditions can influence the sliding contact friction coefficients. Under high loads and at low-interface velocities, composite bearings show minimal rolling friction between the contact surfaces. The coefficient of friction rises with increasing speeds, although under low loads and high-velocity conditions or any other unfavorable condition, such as inadequate surfaces or improper alignment, the coefficient of friction for a composite bearing is low [16].

4.3 STRENGTH AND LOADS

Composite bearings excel in high-static load environments and also perform well under low-speed dynamic loads. Bearing manufacturers typically offer a full spectrum of specialized products that can meet the requirements of each application, whether that be large loads, high speeds, or extreme temperatures. As shown in Table 3.2, composite bearings tend to meet or exceed competitive traditional designs in most bearing metrics, especially in maximum dynamic bearing capacity. Almost all bearing designs offered commercially by GGB by Timken have higher maximum dynamic bearing capacities than traditional materials such as bronze and aluminum [14]. Only hardened steel is stronger than the composite material, but it is important to note that steel bearings require lubrication and are not recommended for corrosive surroundings. Composite materials do not require any external lubrication and will not corrode, even within the most reactive environments. The bearing performance given in Table 3.2 limits the maximum dynamic bearing capacities to only be applicable below 0.025 *m/s* [14]. Composite bearings struggle at high rotational speeds, and many

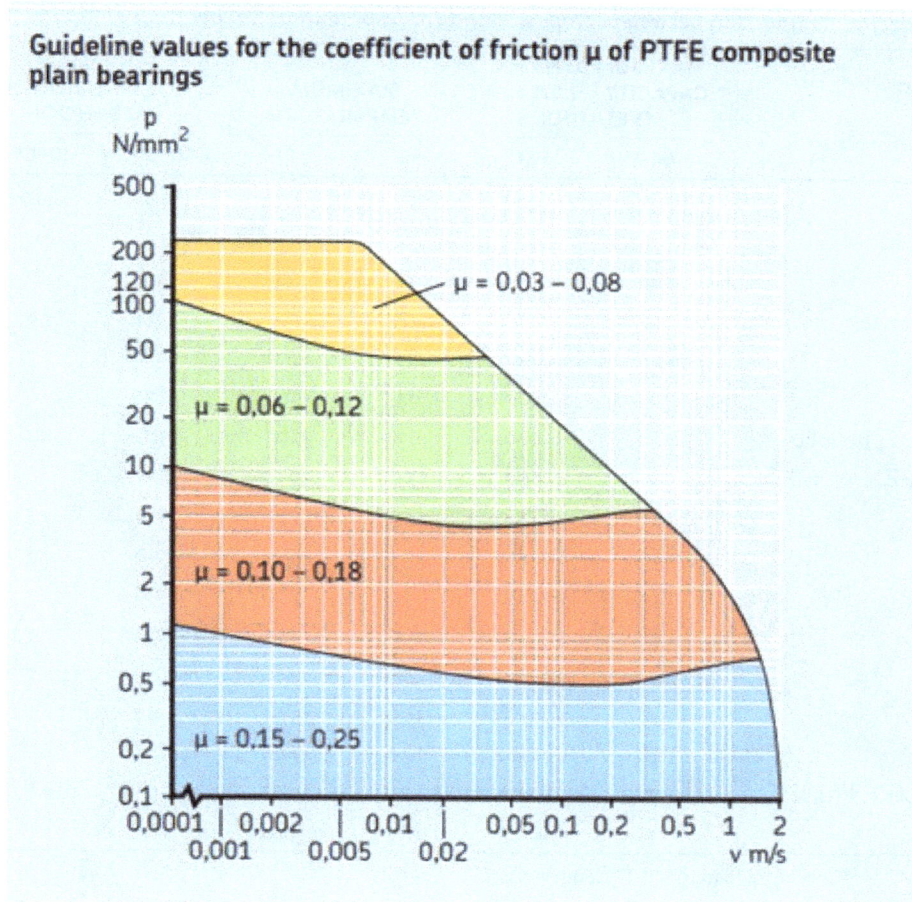

FIGURE 3.6. Operating conditions and coefficient of friction [16].

manufacturers only offer bearings rated for speeds below specific limits, as shown in the product comparison in Table 3.3 [14].

This limitation on rated speeds may seem restrictive, but there remain countless potential applications where composite bearings could be helpful, particularly in the realm of off-road vehicles. Composite bearings are not suitable for applications involving high-speed shafts, such as crankshafts in an engine and axle shafts in an automotive differential, but they can be an excellent choice for low-speed cyclic loads such as those found in an excavator arm and a power steering mechanism.

The combination of load and speed is an important consideration to a designer, and this metric is the primary factor in approximating bearing life. An engineer will want to work closely with potential bearing suppliers and provide them with appropriate values for use in their design calculations. Generally, it is important to minimize the specific load and sliding contact speed to maximize bearing life. It is also important to size the bearing appropriately. Increasing the bearing length will reduce the specific load but will also introduce additional and often significant new loads if the shaft deflects inside the bearing [14]. There can be contradictory requirements in design, and compromises must often be made to best fit individual applications with appropriate choice in composite bearings.

TABLE 3.2. *Properties comparison between composite and traditional material bearings [14]*

MATERIAL	MAXIMUM DYNAMIC CAPACITY [<0.025 M/S (5 FT/MIN)]		MAXIMUM TEMPERATURE		THERMAL EXPANSION RATE–HOOP		SPECIFIC GRAVITY
	N/MM²	PSI	°C	°F	10–6/K	10–6/°F	
Cast bronze*	41	6,000	71	160	18	10	8.80
Porous bronze**	28	4,000	71	160	18	10	7.50
Alloyed bronze*	69	10,000	93	200	28.8	16	8.10
Steel-backed bronze*	24	3,500	93	200	14.4	8	8.00
Hardened steel*	276	40,000	93	200	12.6	7	7.90
Zinc aluminum*	38	5,500	93	200	27	15	5.00
Fabric-reinforced phenolic*	41	6,000	93	200	36	20	1.60
Reinforced PTFE	14	2,000	260	500	99	55	2.00
GAR-MAX®	140	20,000	160	320	12.6	7	1.87
GAR-FIL®	140	20,000	205	400	12.6	7	1.96
HSG	140	20,000	160	320	12.6	7	1.87
MLG	140	20,000	160	325	12.6	7	1.87
HPM	40	20,000	160	320	12.6	7	1.87
HPMB®	140	20,000	160	325	12.6	7	1.87
HPF®, sliding plate	140	20,000	140	285	10.8***	6***	1.90
GGB MEGALIFETM XT	140	20,000	175	350	12.6***	7***	1.85
MULTIFIL	35	5,000	280	540	—	—	2.37

*With lubrication   **Oil impregnated   ***Lengthwise

4.4 THERMAL PROPERTIES

Composite bearings provide impressive temperature resistance and, in some cases, better thermal expansion rates than traditional bearing materials. Many traditional bearing materials, such as bronze and steel, require lubrication to function properly. Even if the bearing material itself may be able to sustain higher temperatures, oils and greases will quickly deteriorate at high temperatures. Composite bearings, as shown in Table 3.2, generally allow for higher maximum sustained temperatures than traditional materials. They are typically needed in assemblies that see elevated operating temperatures above 90°C. High temperatures are a limiting factor on applicability, as the bearing surface will soften under prolonged exposure to extreme thermal stresses above 150°C [14]. In some cases the composite bearing may still function, but it will have a lower load capacity and a shorter lifespan. In fact, as detailed in Figure 3.7, a decreased expected bearing life begins at 65°C [14]. Even though composite bearings have high maximum temperatures capabilities, sustained high temperatures are not a preferred operational environment, and elevated temperatures will have a detrimental effect on bearing life.

Composite bearings also show excellent thermal expansion rates. For a bearing, thermal expansion in the "hoop" direction should be minimized as much as possible. Thermal expansion can place additional stresses

TABLE 3.3. *Properties of GGB bearings [14]*

PHYSICAL PROPERTIES	UNITS	GAR-MAX®	GAR-FIL®	HSG	MLG	HPM	HPMB®	HPF®, SLIDING PLATE	GGB MEGALIFETM XL	MULTIFIL
Ultimate compressive strength	N/mm²	414	379	621	414	345	414**	379	207	—
	psi	60,000	55,000	90,000	60,000	50,000	60,000	55,000	30,000	—
Static load capacity	N/mm²	210	140	415	210	210	210	140	140	70
	psi	30,000	20,000	60,000	30,000	20,000	30,000	20,000	20,000	10,000
Maximum dynamic load capacity	N/mm²	140	140	140	140	140	140	140	140	35
	psi	20,000	20,000	20,000	20,000	20,000	20,000	20,000	20,000	5,000
Maximum relative surface speed	m/s	0.13	2.50	0.13	0.13	0.13	0.13	2.50	0.50	2.50
	fpm	25	500	25	25	25	25	500	100	500
Maximum pU factor	N/mm² x m/s	1.05	1.23	1.05	1.05	1.23	1.23	1.23	1.23	0.32
	psi x fpm	30,000	35,000	30,000	30,000	35,000	35,000	35,000	35,000	9,000
Maximum operating temperature	°C	160	205	160	160	160	160	140	175	280
	°F	320	400	320	320	320	320	285	350	540
Minimum operating temperature	°C	−195	−195	−195	−195	−195	−195	−195	−195	−200
	°F	−320	−320	−320	−320	−320	−320	−320	−320	−330
Thermal expansion rate—hoop	10–6/K	12.6	12.6	12.6	12.6	12.6	12.6	10.8*	12.6*	—
	10–6/°F	7.0	7.0	7.0	7.0	7.0	7.0	6.0*	7.0*	—
Thermal expansion rate—axial	10–6/K	27.0	27.0	27.0	27.0	27.0	27.0	—	—	—
	10–6/°F	15.0	15.0	15.0	15.0	15.0	15.0	—	—	—
Specific gravity	—	1.87	1.96	1.87	1.87	1.87	1.87	1.90	1.85	2.37

* Lengthwise
** For details, contact GGB Applications Engineering Department

on the shaft, increasing friction and generating additional heat, leading to continued thermal expansion. This cycle feeds on itself continuously, creating accelerated wear and eventual bearing destruction. Table 3.2 compares the thermal expansion of many composite bearings with traditional bearing materials. The composite offerings meet or exceed the traditional materials in this metric. Additionally, Table 3.3 clearly illustrates that composite bearings have a greater thermal expansion capability in the axial direction.

Composite bearings are offered in similar sizes as traditional bearings to work with standard shafts. Composite bearing selection follows the same systematic design process as metallic bearing selection. Manufacturers provide catalogs that include critical dimensions and part drawings for their various bearing selections. Important dimensions to consider are the inside diameter (ID), the outside diameter (OD), and the bearing length (w). The bearing manufacturer will also provide tolerances and clearances for the bearing housing. ID tolerances are critical to ensuring that the bearing fitment is suitable for the application without being too tight or too loose. In the past, composite material-based bearings have been limited in applicability by the inability to provide tightly toleranced parts due to the composite material manufacturing processes and filament

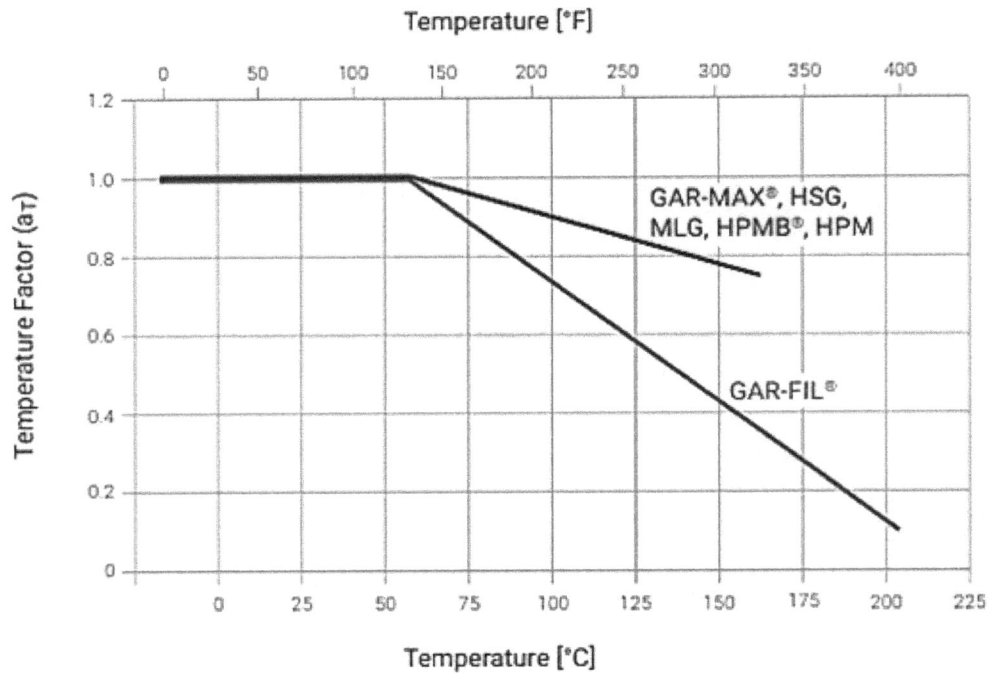

FIGURE 3.7. Temperature factors for bearing life calculations [14].

winding processes used to produce composite bearings. Most metallic bearings are machined from steel. The ability to apply modern high-precision machining processes allows metallic bearings to be manufactured with extremely tight tolerances. Composite bearings are made in an additive manner by winding filaments and resins rather than removing material to size with a computer numerical controlled (CNC) machine tool. Previously this introduced uncertainty into the composite manufacturing process, but advancements in fiber construction for self-lubricating materials used in the assembly of composite bearings have allowed the production of bearings with similar dimensional tolerances to metallic bearings [20].

Composite bearing manufacturers are very transparent about their specified tolerances. The specifications outlined in product catalogs and design guides include ID and OD tolerances, recommended bearing housing bore diameters, press-fit tolerances, pin diameter recommendations for sleeve bearings, running clearances, and bearing length tolerances. Table 3.4 presents an example of a typical design guide used for composite bearing selection [18].

Proper sizing calculations and load considerations must be made to achieve optimal bearing performance. Bearing sizes are driven by the radial forces acting on the shaft. When determining the radial and axial forces being applied to the shaft, there are three main considerations: static loads, applied loads, and drive loads. Static loads are most commonly loads produced by the weight of the components. Applied loads are those derived from normal operating conditions on the component, such as thrust loads exerted onto a shaft. Drive loads are generally torques transmitted by the power source to the load. Although there are many steps to ensure that a bearing is properly sized, the calculation of radial force for belt-,

TABLE 3.4. *Bearing selection charts [18]*

BEARING PART NUMBER	NOMINAL ID (IN)	ID (IN)	OD (IN)	RECOMM-ENDED HOUSING BORE (IN)	PRESS FIT (IN)	RECOMM-ENDED PIN DIAMETER (IN)	RUNNING CLEARANCE (IN)	LENGTH TOLERANCE (IN)
PWR08M16	1/2	0.5058	1.0025	1.000	0.0005	0.5000	0.0013	+/-0.010
		0.5088	1.0045	1.002	0.0045	0.4980	0.0103	
PWR10M18	5/8	0.6309	1.1276	1.125	0.0005	0.6250	0.0013	+/-0.010
		0.6339	1.1296	1.127	0.0046	0.6230	0.0104	
PWR12M20	3/4	0.7560	1.2526	1.250	0.0006	0.7500	0.0014	+/-0.010
		0.7590	1.2546	1.252	0.0046	0.7480	0.0104	
PWR14M22	7/8	0.8811	1.3777	1.375	0.0006	0.8750	0.0014	+/-0.010
		0.8841	1.3797	1.377	0.0047	0.8730	0.0105	
PWR16M24	1	1.0065	1.5027	1.500	0.0007	1.0000	0.0015	+/-0.010
		1.0092	1.5047	1.502	0.0047	0.9980	0.0105	
PWR18M26	1–1/8	1.1314	1.6278	1.625	0.0007	1.1250	0.0016	+/-0.010
		1.1344	1.6298	1.627	0.0048	1.1230	0.0107	
PWR20M28	1–1/4	1.2564	1.7528	1.750	0.0008	1.2500	0.0016	+/-0.010
		1.2594	1.7548	1.752	0.0048	1.2480	0.0106	
PWR22M30	1–3/8	1.3816	1.8779	1.875	0.0008	1.3750	0.0017	+/-0.010
		1.3856	1.8799	1.877	0.0049	1.3730	0.0117	

chain-, and gear-driven shafts is the same for both metallic and composite bearings [1] and is presented as Equation #1:

$$F_r = \frac{(P)(19.1 \times 10^6)(K_t)}{(PD)(\Omega)}$$ (1)

where

F_r = radial force on shaft (*N*)
P = power transmitted (*kW*)
PD = pitch diameter (*mm*)
Ω = shaft speed (*rpm*)
K_t = drive tension factor (1 for chains & gears, 1.5 for belt drives)

5. COMPOSITE BEARINGS IN HEAVY OFF-ROAD EQUIPMENT

Off-road equipment design provides numerous constraints and criteria that could be interpreted as advantageous toward the use of composite bearings by an off-road vehicle manufacturer. These bearings minimize the need for continuing maintenance as the machinery is operated. They provide a positive effect on the field capacity of the equipment and are beneficial to the overall life of the equipment in corrosive and abrasive environments. A few illustrative case studies have been provided, showing the potential use of composite bearings in the design of agricultural, construction, forestry, and mining vehicles.

5.1 REDUCTION OF MAINTENANCE

Off-road vehicles require regular service on traditional metallic bearings at specific operating intervals to keep the equipment operating at its optimal performance. These service intervals include grease applications for the multitude of bearings and other moving mechanical components. Grease intervals can range from daily to seasonal, and it is now standard to label the grease fitting location with a specific decal featuring the recommended number of operational hours between servicing. The heavy equipment industry uses National Lubricating Grease Institute (NLGI) #2 grease for most greaseable parts, unless otherwise specified. Grease is rated with an NLGI number, which represents the consistency of the lubrication and is derived using the American Society for Testing Materials (ASTM) standard D217 [24]. This standard outlines the testing methods for the cone penetration of lubricating grease at $25°C$ [25]. The grease for use on the operating equipment must be purchased, and labor must be allotted for the lubrication process. Replacing traditional metallic bearings with composite bearings eliminates both of these expenses for the converted point of application.

5.2 COST-EFFECTIVENESS

Regular servicing is time-consuming for operators to complete but is a necessary step for continuing trouble-free daily operation. Minimizing the amount of grease-requiring components on off-road vehicles would be highly beneficial for a machine's field capacity, but it might not be feasible for every specific function on a machine. However, replacing traditional metallic, grease-requiring bearings with greaseless composite bearings could eliminate the need for daily service in many locations and would allow more time per shift for equipment productivity. The cost of grease can be a sizable annual maintenance cost for a variety of different industries, especially for those that are managing large fleets of equipment. Designing equipment with more self-lubricating composite bearings could significantly reduce the continuing costs of ownership and help minimize equipment downtime for maintenance.

Vehicle maintenance typically falls under the vague accounting charges of indirect labor and miscellaneous supplies. To illustrate the somewhat hidden expense of grease and routine equipment service in maintenance budgets, calculations were performed for the cost to service all 25 h grease fittings on a standard Class VIII combine for one harvest season in the midwestern region of the United States. An axial combine used for corn, soybean, and cereal grain harvest in the Midwest has approximately 20 grease fittings around the machine, requiring service every 25 h of engine operation. Additionally, there are around seven grease fittings that require service every machine operating day, along with numerous others having different intervals. The number of grease fittings on combines vary between manufacturers, but they are all comparable. Each requires routine maintenance that creates downtime for the machine. Most grease fittings need three to five pumps of grease per fitting unless otherwise specified by the manufacturer. A standard, manual-lever grease gun supplies about 1.3 g of grease per pump [21]. Harvest seasons in the midwestern United States are typically around eight weeks long. It was assumed that 75% of the total days for the completion of harvest season are combine-operating

TABLE 3.5. *Inputs for Grease Cost Analysis Calculations*

Qty. of 25 *h* grease fittings	20	fittings
Flow rate/pump	1.3	*g*
Avg. qty. of pumps/fitting	4	pumps
Avg. length of harvest	56	days
Estimated combine operating days	42	days

days, since it is not reasonable to assume that a combine would operate continuously through the length of the harvest season. It was also assumed that a 25 h combine operating interval is equivalent to approximately two *days*. Table 3.5 and Equations 2, 3, and 4 estimate the cost of grease to service one standard combine for a single harvest season. Ten standard 14 *oz* tubes of #2 NIGL multipurpose grease were used to calculate the average price used in Equation 2. The cost of the 25 h service interval for the harvest season was roughly US$60.

$$Total\ Grease\ for\ One, 25\ h\ Service = 20\ (1.3\ g * 4\ pumps) = 104\ g \tag{2}$$

$$Total\ Cost\ of\ One, 25\ h\ Service = \left(\frac{104\ g}{14\ oz.}\right) * \left(\frac{1\ oz}{28.3\ g}\right) * \$10.98 = \$2.82 \tag{3}$$

$$Total\ Cost\ for\ One\ Harvest\ Season = \$2.82\left(\frac{42}{2}\right) = \$59.50 \tag{4}$$

Service cost estimations become proportionally larger for agricultural and construction operations with multiple pieces of equipment and various implements requiring regular servicing within their fleets. Table 3.5 and Equations 2–4 are for a single machine and a single specific service interval. In addition to the 25 *h* service interval, combines will typically also have 10 *h*, 50 *h*, 75 *h*, 100 *h*, and 200 *h* grease fitting service intervals, with their own annual charges. Maintenance neglect at any of these service intervals will cause a premature failure of bearings and potentially other larger mechanical components. The total service cost for each unit would be the sum of the service intervals, and the total cost to an entity would be multiplied by the number of vehicles in their operational fleet.

5.3 RESISTANCE TO HARSH ENVIRONMENTS

Providing resistance to resist debris entrance and contamination is one of the greatest challenges to overcome when designing heavy equipment for the agricultural, construction, forestry, and mining industries. Engines, mechanical motion parts, fluid power, and electronics are highly sensitive to dust, debris, and other contaminants found in the environments where heavy equipment spends most of its operational life. While metallic bearings and composite bearings both should be protected from contamination, neglecting bearing seal integrity for metallic bearings will cause a premature failure at a more rapid rate in these types of environments. The threat of bearing failure caused by contamination and lack of service is minimal for a self-lubricating composite bearing. This feature also makes composite bearings an excellent choice in difficult-to-access locations. Some bearings on heavy equipment require the removal of several parts to gain access to grease fittings for service. This often causes operators to not apply the proper amount of grease to the bearing or ignore its required service intervals altogether. A common solution to this problem is the installation of remote grease fittings on a bulkhead in an accessible location, connected to the bearing housings with steel lines or hoses. However, remote grease fittings require more grease initially because the lines must be filled with grease. Hard-to-reach bearing locations outfitted with self-lubricating composite bearings are an excellent solution to issues with impeded service accessibility and the potential for premature bearing failure if neglected.

5.4 CASE STUDIES

Some real-world equipment issues that were resolved through the application of composite bearings in heavy equipment designs are summarized in this section. Many of these solutions included maintenance reductions, increased wear resistance in corrosive environments, and economic advantages in both design and manufacturing. These case studies were published by Polygon Composites to summarize collaborations between Polygon and various equipment manufacturers looking to resolve problematic designs deriving from metallic bearings [26].

5.4.1 LIVESTOCK FEED MIXER

Polygon and Kuhne Industrie BV, a Netherlands-based manufacturer of composite bearing materials, collaborated to produce a custom composite bearing to work with an agricultural equipment manufacturer's tandem-axle, feed-mixing cart. The equipment manufacturer had tested several bearing types, all of which caused excessive friction and premature wear on the shafts due to the high and uneven load on the bearings during operation. In addition to not supporting the known load demands of the equipment, properly sized bearings were difficult to procure. Polygon and Kuhne worked closely to design a composite material solution for the feed-mixing cart manufacturer. This led to the development of a custom composite bearing that withstood the high and uneven loads, minimized the need for maintenance, and reduced noise caused by the rapid oscillation of the assembly held by the bearing [27].

5.4.2 REDUCED MAINTENANCE IN BULLDOZER DESIGNS

Polygon provided composite bearings and bushings to three major US construction equipment manufacturers as a greaseless bearing alternative that minimized chassis maintenance requirements. Bearing service in heavy equipment applications can be very time-consuming and difficult depending on the location of the bearing to be serviced. Due to these common challenges, one of the leading bulldozer manufacturers in the United States approached Polygon for a composite bushing to replace a traditional, metallic bushing located in a position that was difficult for operators and maintenance technicians to service with grease. This bushing was designed as part of an assembly that connected the bulldozer track assembly to the machine's chassis. To solve this critical design challenge, Polygon developed a self-lubricating composite bushing with an industrial wear surface that permanently mated with the internal diameter of the bushing. Due to the development of this wear surface, this bearing does not need to be greased or serviced for the design life of the equipment and will not corrode due to the nonmetallic construction of the bushing. Since this initial technology introduction, the utilization of this composite bushing has become essential in bulldozer designs by all major US construction equipment manufacturers [28].

5.4.3 COMPOSITE BUSHINGS IN EARTHMOVING EQUIPMENT

Polygon has designed composite bushings for K-Tec earthmoving equipment, a Canadian-based supplier of mining equipment. K-Tec used Polygon's bushings in its hydraulic cylinders and for other critical mechanical pivot points on its machinery. Mining equipment is operated in extremely harsh environments, with many contaminants and debris present in the ambient air, water, and soil of the mine. Equipment running in this industry must be rugged and able to withstand high loads and must perform reliably with minimal maintenance. Leveling and scraping equipment uses hydraulic actuation to adjust the depth of the working edge for various digging and scraping duties. These hydraulic leveling systems must withstand high amounts of strain and stress and must operate accurately on a daily basis. To reduce the need for daily lubrication service, Polygon developed custom greaseless bushings for the hydraulic linkage pivot points that can withstand the severe-duty requirements of the equipment and maintain functionality in the debris- and contaminate-laden environment of the mines [29].

6. TESTING AND VALIDATION

Reputable manufacturers of composite bearing products complete rigorous testing to ensure their products' capabilities and evaluate them against other competitors' products. Common bearing tests include rotational wear, linear wear, pivoting wear, and friction coefficient measurements. Igus has published several of its product experiments as well as the test reports completed for its iglide® composite bearing line [19]. Igus's test lab has researched low load pivot wear, composite bearings versus metal bearings on different shaft types, frictional wear on composite bearings over many cycles, and plain bearing and brass bearing wear on various shafts [30]. In one circumstance, several variations of iglide® composite bearings were tested against a customer's brass bearing, which was used on a rotating shaft. During operation, the brass sleeve bearing was subject to "tumbling" load stresses. This term describes a shaft that wobbles, creating uneven wear patterns. The shaft does not wear uniformly, due to the tumbling effect, and typical metallic bearings in these applications have low lives. Igus tested two different variations of composite bearings, each with two different lubrication configurations, along with two different configurations of a competitor's brass bearing. All experimental components were tested with and without grease. The purpose of the testing was to discover the optimal bearing material for the specific intended application that reduced both shaft and bearing wear over time. It was demonstrated that an average of 490.4 $\mu m/km$ of bearing wear was measured for the dry-running iglide® bearings, compared to the measured wear of 1,325.0 $\mu m/km$ for dry-running brass bearings. An average of 425.4 $\mu m/km$ of bearing wear was measured for the grease-applied iglide® bearings, compared to 497.6 $\mu m/km$ of measured wear for the grease-applied brass bearing. However, the most optimal bearing, when subjected to the unique tumbling stress was the grease-applied iglide® Q bearing, which showed the least amount of measured wear at 241.2 $\mu m/km$, compared to the customer's brass bearing [30]. The results for each bearing type and lubricant configuration are shown in Figure 3.8 [30].

Composite bearing manufacturers typically use state-of-the-art testing facilities to validate their products' performance. Obtaining high-confidence testing data for a specific application is critical for product design and allows for comparisons to be drawn between the product and other competitive products for marketing purposes. These tests are designed to simulate the working conditions of the equipment in a variety of applications and are primarily conducted to measure wear and friction. The design guides that many bearing manufacturers provide can help determine the optimal bearing for an intended design and application. Specific results with the data collected during well-designed product testing and validation program would be needed prior to any full commitment to convert a design to composite bearings.

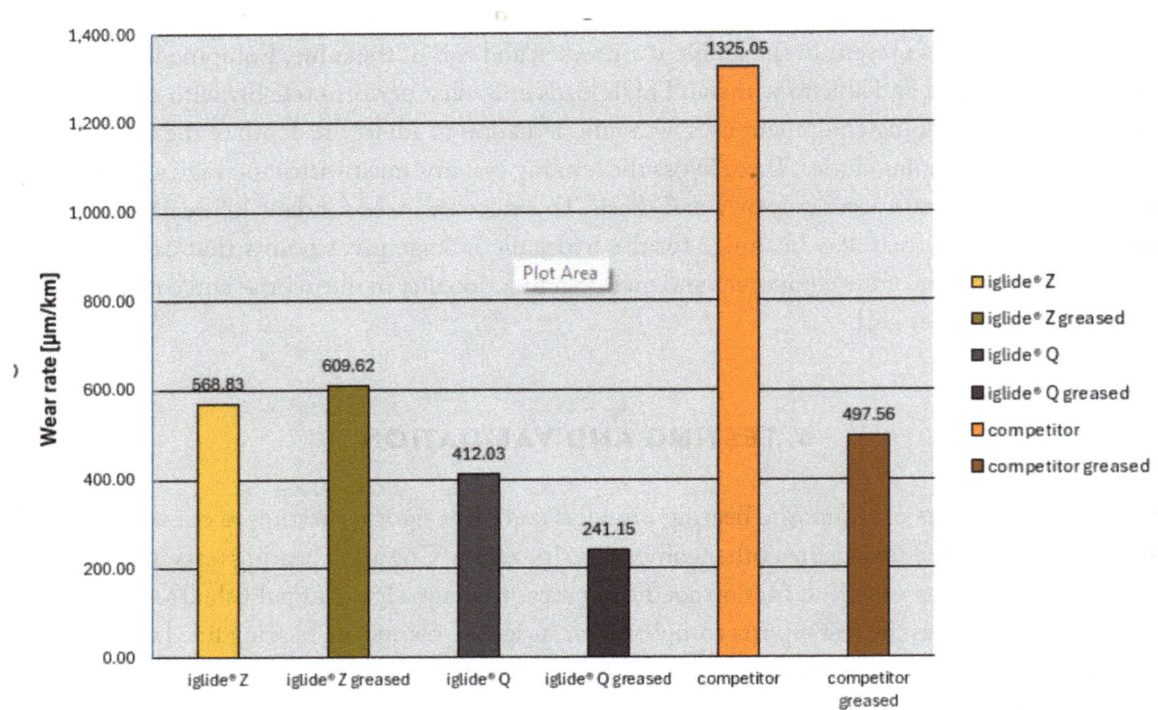

FIGURE 3.8. Measured wear ($\mu m/km$) (y-axis) test results for iglide® composite bearing in tumbling motion [30].

7. DESIGN CONSIDERATIONS

The overall design of a composite bearing joint proceeds exactly like a traditional metallic bearing joint [1]. All motions and loads must be accurately categorized, and performance criteria should be specified. Any specific constraints for the application should be noted. Design guides from specific manufacturers will now be useful. It is always wise for engineers to explore their own solutions with a design guide prior to contacting a company to familiarize themselves with the company's key parameters and processes. It may be useful to utilize multiple potential vendors. After an introduction, a prototype product for the specific application should be developed as a joint project. Testing and validation should be planned and reviewed before committing to a long-term partnership. These steps will be explored in the following sections in greater detail.

7.4 DESIGN GUIDES

Design guides provide engineers and technologists with the details and specifications necessary to evaluate which products are best suited for the specific design. These guides are published by bearing manufacturers and are available online. Often there is an abundance of information in these product design guides, which should be appreciated by modern engineers and technologists. Obtaining critical details regarding a product can sometimes be difficult due to proprietary constraints and lack of transparency about design. Bearing design guides contain specifications on mechanical properties, thermal properties, dimensions, and sizing

CALCULATING SLEEVE BEARING PV LIMIT

Example: .750" Pin Dia @200 rpm , 85.0 lb. total load, bearing length .750"

A = .750 (Pin) x .750 (bearing length) = .562 in.^{2}

V = 0.262* x rpm x diameter
= 0.262 x 200 x .750 = 39.3 ft/min

P = 85.0 lbs. / .562 in.^{2}
= 151.2 psi

P = total load / projected area (A)**

PV = 151.2 psi x 39.3 ft/min
= 5,942 psi* ft/min

FIGURE 3.9. Pressure and volume limit calculations [15].

recommendations for working stresses. Different bearing classifications relative to the direction of motion in the intended application may also be included. Many of these design guides also incorporate specific design calculations to aid with bearing sizing and selection within the manufacturer's product line. Figure 3.8 provides calculations and recommendations for pressure and velocity in Polygon's bearing selection and design guide [15]. Figure 3.9 shows the physical forces and geometry of the bearing defined in Polygon's preferred terminology [15]. The sample example calculation uses a 1.9 *cm* (0.75 *in*) diameter pin turning at 200 *rpm* and supported by a sleeve bearing subject to 380 *N* (85 *lb*$_f$) of load. The pin diameter and rotational speed of the pin are utilized in the velocity calculation, while the load on the bearing over the projected contact area is used in the bearing pressure calculation. The solutions to P and V are multiplied to equate a singular, quantified metric for comparison against other sleeve-type bearings. Pressure and velocity limit calculations are standard methods for comparing bearing performance [15]. Design engineers use PV calculations to determine safety limits with respect to loading and operating speed for the application.

7.5 BASIC DESIGN AND MANUFACTURING

There are standard design practices associated with the design and manufacturing of all products. Bearing assemblies are no different. For composite bearings, some design practices that should be utilized are to ensure the correct material choice for the bearing, press-fit interference tolerance in the assembly, minimized post-curing effects in the backing layers, proper overall length, and correct surface finish. Bearing material selection is important, because most composite materials have a greater thermal expansion coefficient than metallic bearings and will require larger clearances to properly fit across all operational temperatures. Proper mating materials in the assembly are also very important, because a harder mating material can result in a longer bearing life [31]. Interference tolerances are important in bearing design, because proper secure fitment is critical to bearing life. During the post-curation process for the composite material, it is recommended the bearing be baked at 140–150°C for four hours or annealed in SAE 20 oil at 107–130°C [31]. This provides the best material properties for long-term bearing life. A best design practice regarding the length of the bearing is for the L:D ratio to not exceed 3.5 [31]. Unexpected forces in a bearing can develop on rotating joints with slender pins. The composite bearing surface finish should be molded if possible to provide the least amount of wear in motion and the lowest possible coefficient of friction [31]. These common design best practices have been listed to provide a general understanding of the variety of different processes that manufacturers use to measure and control to provide critical performance properties in composite bearings.

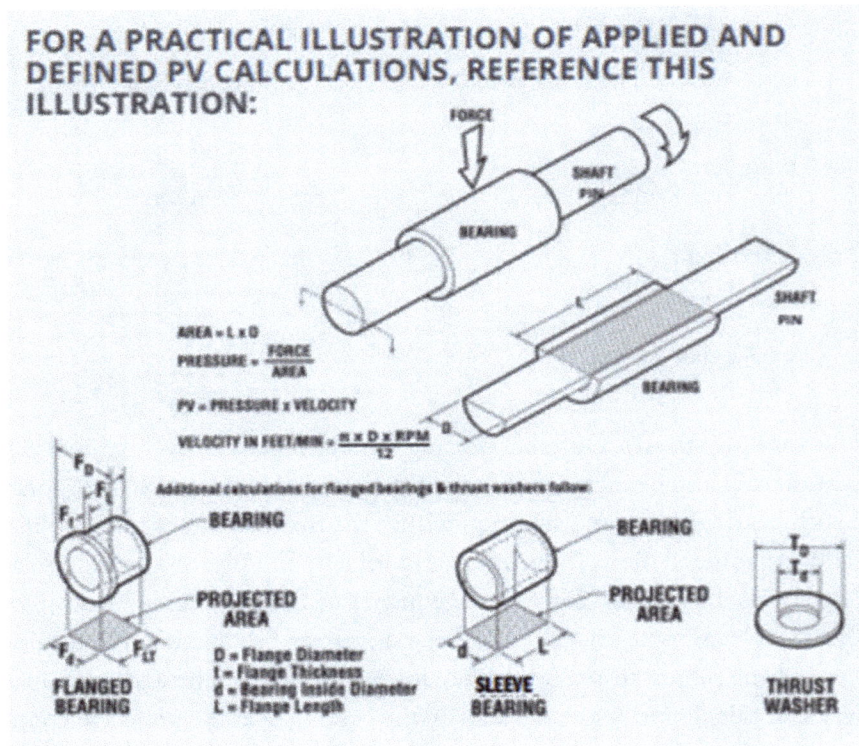

FIGURE 3.10. Geometry and forces present in sleeve bearing calculations [15].

7.6 PRODUCT AVAILABILITY AND PROCUREMENT

When an off-road vehicle is being designed, parts that are not produced in-house or are being prototyped require vendor sourcing. Vendor relationships and shared business processes are critical for the successful manufacturing and maintenance of composite bearing systems in sustained heavy-duty applications. The collaborative themes discussed previously will be expanded in this section.

The three ways to determine the necessary product fitment for components in the manufacture of heavy-duty off-road-vehicles are ordering existing parts from a manufacturer of the desired component, purchasing mass-produced parts from an authorized retailer, and partnering with a company to create a new specific product for use through joint research and development efforts. When purchasing existing components, it is generally advised for engineers to look for vendors that already supply products to their company due to the history of past transactions and an established business partner relationship. If these vendors do not have the needed product and cannot or will not help produce a new product to fit the specifications, then it is advisable to seek new vendors. This can be an involved process, since a new business-to-business relationship is being developed as opposed to a technical one.

Existing products are manufactured items that have a history of usage behind them and are generally available in large quantities due to their established demand. Many times these products can be found through wholesalers or parts houses. These products typically have extensive testing and engineering support proving that the product works as claimed, which means these types of components are good selections for use in a production design. When an engineer seeks a new product, research on the manufacturer or vendors is critical to determining if the desired product from this particular vendor is suitable for use. For example, one can go to the iglide® website, select the "Plain Bearings" tab, and search for a component that would fit most products' specifications. Many if not all commercial websites have filter applications available for easy site navigation, and it is advisable to utilize these useful tools when selecting components for a larger design.

If there are no products on the market that are suitable for a specific design, then one must consider manufacturing a new part to complete the assembly. Generally, it is more cost-effective and time-efficient to contact a manufacturer already in the business to have a prototype designed for the specific new assembly rather than attempt to make one in-house. If the component is something that is related to a trade secret or is in need of intellectual property protection, then the necessary legal protections must be in place before divulging any information to another corporation. If these protections are needed and adequately secured, then a manufacturer who deals in the desired product area can be contacted. These organizations are typically more than happy to work with customers and assist them in finding the right part for the job. New design prototypes will cost more than an existing part, due to the need to create new tooling, casting fixtures, and other equipment to manufacture test batches before full production can be made. However, this is a onetime cost. If the new prototypes are to be produced for a long period of time, then the added costs will be negligible, amortized over the entire production run.

8. CONCLUSION

The advantages, unique properties, and design characteristics of composite material bearings have been illustrated by this design overview in an effort to convey the state of the art for this technology to practicing engineers and technologists working on all types of designs for off-road heavy equipment. The off-road vehicle industry should be able to take advantage of many types of composite bearings to incorporate near-frictionless, maintenance-free, and contaminate-resistant motion into their equipment designs. Metallic bearings, which use grease or oil to create a near-frictionless barrier between mechanical motion parts, are preferred in many heavy equipment applications, but these components have continuing financial and time costs associated with their use. Traditional metallic bearings feature high load factors, durability, and low operating noise. Metallic bearings function very well when maintained properly in accordance with specific design specifications and recommended service intervals, which typically require the application of grease or oil. Maintenance is critical to bearing life and avoidance of premature bearing failure. Composite bearings maintain critical functions and reduce the need for continuing maintenance.

Servicing every day or once per week may seem like a minute challenge. However, for large commercial construction, forestry, mining, and agricultural operations, maintenance is very costly in two primary ways. The cost of consumables, tools, and labor adds significant cost to properly maintaining equipment. A numerical experiment on a modern combine, where the costs of grease and labor were evaluated for only one specific service interval, demonstrated the potential of these expenses to grow and become significant. For many machines, there are often five to seven different service interval periods with many different grease fittings and associated intervals. The additional costs that stem from equipment downtime during the maintenance period reduce the profitability of the machinery. Equipment downtime is detrimental to machine efficiency, field effectiveness, and return on investment. These factors are always analyzed in detail by equipment and project managers to accurately predict operating costs, making a composite bearing–equipped machine a potential superior consumer choice. Another design challenge that might encourage composite bearing use includes accessibility issues for a specific bearing location. Off-road equipment utilizes many complex chassis designs with tough-to-reach areas. Many of these inaccessible locations have bearings that require regular service or occasional replacement. Sometimes a remote grease fitting is not an option, depending on the bearing type. Additionally, some heavy equipment operates in corrosive environments that can attack metallic bearings. Composite bearings can be used to increase the useful life and maintenance intervals of such vehicles.

Composite material bearings offer a variety of solutions to each of these unusual challenges. Case studies were reviewed on the application of composite bearings to unique situations, where metallic bearings were problematic or were not an option. A leading bulldozer manufacturer significantly reduced downtime for maintenance through the integration of a composite bushing that did not require regular lubrication to maintain optimal bearing performance and would not corrode in harsh environmental conditions. This improvement was provided by utilizing a bearing that could last the design life of the equipment. This is a prime example of the advantages associated with composite bearings in heavy-duty off-road vehicles. The self-lubricating features provided by PTFE and other special coatings eliminate the need for applied lubrication and regular access. Additionally, composite resins and materials are non-corrosive, making them extremely advantageous in adverse conditions, where many types of off-road equipment spend most of their life cycles.

Many types of bearings and configurations are applied in many types of machinery designs to provide a frictionless barrier, smooth motion, and minimal operating noise for working machine parts. The heavy-equipment industry, specifically the agricultural, construction, forestry, and mining segments, could potentially integrate many composite bearings into their designs for the basic vehicles and auxiliary implement functions on their machines. The specific heavy-duty mechanical components designed for specific tasks and operations differentiate off-road equipment from the automotive industry and other on-road equipment. Composite bearings will allow these machines and chassis designs to improve through the elimination of accessibility and continuing service requirements, providing significant cost reductions for consumers and equipment managers.

9. RECOMMENDATIONS

The scope of this chapter was to introduce composite bearing technologies and educate design engineers and technicians on the advantages of using non-traditional composite bearing types in mobile and industrial equipment design applications. Emphasis was placed on the off-road vehicle industry in the agricultural, construction, mining, and forestry equipment manufacturing sectors, given the off-road industry's unique ability to take advantage of the anti-corrosive and maintenance-free characteristics of the composite bearings. There are numerous other reasons and advantages to incorporate composite bearings into designs. It is recommended that designers consider composite bearings in their research and development processes by reviewing and utilizing the detailed design guides provided by several composite technology manufacturers, some of which were described. It should be expected that composite bearing modules will appear in computer-aided design software soon to assist in embedding these new bearings in different designs. Similar to the ability to import any component data from a supply house for use in a new design, this inclusion will speed up the introduction of this technology into common use.

Many composite technology companies have exceptional technical service engineers and are willing to create custom solutions for incorporation into a variety of design applications. The emphasis on custom bearing applications is greater in the composite technology industry than with traditional metallic bearing manufacturers. Many composite bearing manufacturers have a systematic engineering process for developing custom solutions for unique applications requiring their product. This begins with a preliminary study and direct conversations with a client to formulate the background and scope of the problem. This initial step is typically followed by a joint design process that eventually leads to product validation and testing. Once testing has been completed and a solution has been optimized for a specific application, mass production can take place for the inclusion of the composite bearing into a new off-road vehicle design. There are many resources available to take advantage of in the composite bearing industry, and the product transparency offered by these companies will be greatly appreciated by engineers and technicians looking to take advantage of these innovative products.

10. ACKNOWLEDGMENTS

The fall 2023 Design of Off-Road Vehicles class at Purdue University's School of Agricultural and Biological Engineering is acknowledged for their contributions to the structure and content of this technical chapter. Dr. Carol S. Stwalley is graciously thanked for her editorial work with the manuscript and in the final formatting of the document.

11. COMPOSITE BEARING QUESTIONS

1. Describe how composite bearings differ from traditional anti-friction (rolling contact) bearings and how these differences can be advantageous when operating within contaminant-laden environments.
2. What benefits can be attributed to the use of composite bearings, and how can those benefits be used to economically justify the inclusion of more expensive composite bearing components in an off-road equipment design?
3. Describe the conditions in which it might be advantageous to utilize a composite bearing in agricultural applications, construction applications, forestry applications, and mining applications.
4. Identify differences in the bearing design process needed to utilize composite bearings instead of traditional metallic roller bearings.
5. Describe the circumstances in which composite bearings would not be an appropriate choice for inclusion in an off-road vehicle design.

12. REFERENCES

[1] G. W. Krutz, J. K. Schueller, and P. W. Claar II, *Machine Design for Mobile and Industrial Applications*, Warrendale, PA, Society of Automotive Engineers, Inc, 1994, ISBN 1-56091-389-4.
[2] saVRee, "Plain Bearing," 2023. [Online]. Available: https://savree.com/en/encyclopedia/plain-bearing. [Last accessed December 27, 2023].

[3] M. Forsthoffer, "Journal (Radial) Bearings," in *Forsthoffer's Component Condition Monitoring*, Oxford, UK, Butterworth-Heinemann, 2019, ISBN: 978-0-12-809599-7.

[4] GGB by Timken, "How Does a Self-Lubricating Bearing Work?," 2019. [Online]. Available: https://www.ggbearings.com/en/why-choose-ggb/faq/bearings-faq/what-self-lubricating-bearing. [Last accessed December 27, 2023].

[5] R. F. Gibson, *Principles of Composite Material Mechanics*, New York, NY: McGraw-Hill, 1994, ISBN: 978-1498720694.

[6] P. M. Karandikar, R. R. Kharde, S. B. Bhoyar, and R. L. Kadu, "Study the Tribological Properties of PEEK/PTFE Reinforced with Glass Fibers and Solid Lubricants at Room Temperature," *International Journal of Current Engineering and Technology*, vol. 4, no. 4, pp. 2401–2404, 2014, https://inpressco.com/wp-content/uploads/2014/07/Paper192401-24041.pdf.

[7] H. N. Yu, S. S. Kim, and D. G. Lee, "Optimum Design of Aramid-phenolic/glass-phenolic Composite Journal Bearings," *Composites Part A: Applied Science and Manufacturing*, vol. 40, no. 8, pp. 1186–1191, 2009, https://doi.org/10.1016/j.compositesa.2009.05.001.

[8] D. L. Burris and W. G. Sawyer, "A Low Friction and Ultra Low Wear Rate PEEK/PTFE Composite," *Wear*, vol. 261, no. 3–4, pp. 410–418, 2006, 10.1016/j.wear.2005.12.016.

[9] F. Henninger and K. Friedrich, "Thermoplastic Filament Winding with On-line Impregnation. Part A: Process Technology and Operating Efficiency," *Composites Part A: Applied Science and Manufacturing*, vol. 33, no. 11, pp. 1479–1486, 2002, https://doi.org/10.1016/S1359-835X(02)00135-5.

[10] F. Henninger, J. Hoffmann, and K. Friedrich, "Thermoplastic Filament Winding With On-line Impregnation. Part B: Experimental Study of Processing Parameters," *Composites Part A: Science and Manufacturing*, vol. 33, no. 12, pp. 1684–1695, 2002 https://doi.org/10.1016/S1359-835X(02)00136-7.

[11] G. R. Jadhav, D. B. Zoman, and D. R. Mahajan, "Analysis of Composite Journal Bearing," *International Engineering Research Journal*, pp. 1998–2002, 2015, https://www.ierjournal.org/pupload/mitpgcon/1998-2002.pdf.

[12] G. Jadhav, T. Badguiar, D. B. Zorman, and M. Jadhav, "Tribological Performance Analysis of Composite Materials for Journal Bearing," *International Journal of Modern trends in Engineering and Research*, vol. 3, 2016, https://doi.org/10.21884/ijmter.2016.3017.apwts.

[13] P. Sengsri, M. R. Marsico, and S. Kaewunruen, "Base Isolation Fibre-Reinforced Composite Bearings Using Recycled Rubber," in *IOP Conference Series: Materials Science and Engineering* 603, Prague, 2019, https://doi.org/10.1088/1757-899X/603/2/022060.

[14] GGB by Timken, "Fiber Reinforced Composite Bearing Handbook," 2023. [Online]. Available: https://www.ggbearings.com/sites/default/files/2023-08/GGB-Fiber-Reinforced-Composite-Bearings-High-Load-Self-Lubricating-Bearings-Bushings_0.pdf. [Last accessed December 27, 2023].

[15] Polygon Composites Technology, "Bearings Design Guide," 2023. [Online]. Available: https://polygoncomposites.com/design-guide-bearings/. [Last accessed December 27, 2023].

[16] SKF, "SKF Composite Plain Bearings," 2012. [Online]. Available: https://cdn.skfmediahub.skf.com/api/public/0901d1968o229dfc/pdf_preview_medium/0901d1968o229dfc_pdf_preview_medium.pdf. [Last accessed December 27, 2023].

[17] Igus Motion Plastics, "Igus Engineer's Toolbox Plain Bearing Design Guide," 2023. [Online]. Available: https://toolbox.igus.com/design-guides/iglide-plain-bearings-design-guide. [Last accessed December 27, 2023].

[18] Polygon Composites Technology, "Industries," 2023. [Online]. Available: https://polygoncomposites.com/industries/. [Last accessed December 27, 2023].

[19] Igus Motion Plastics, "Bearings Are Rigorously Tested within the iglide® Test Laboratory," 2023. [Online]. Available: https://www.igus.com/info/ plain-bearings-test-laboratory. [Last accessed December 27, 2023].

[20] GGB by Timken, "Composite Bearing Design with Improved Tribology," GGB, Thorofare, NJ, 2020. [Online]. Available: https://www.ggbearings.com /sites/default/files/inline-files/GGB-Whitepaper-Composite-Bearings-for-Aggressive-Applications.pdf. {Last accessed December 27, 2023].

[21] B. Fitch, "The Grease Gun: Applications, Uses, and Benefits," 2023. [Online]. Available: https://www.machinerylubrication.com/Read/29356/ grease-gun-anatomy. [Last accessed December 27, 2023].

[22] A. Carrera, "How Do Self-Lubricating Bearings Lubricate?," 2023. [Online]. Available: https://www.tstar.com/blog /qa-how-do-self-lubricating-bearings-lubricate. [Last accessed December 27, 2023].

[23] Bushing MFG, "PTFE Bearings," 2023. [Online]. Available: https://bushingmfg.com/ptfe-bearings/. [Last accessed December 27, 2023].

[24] National Lubricating Grease Institute, "Grease Glossary," 2017. [Online]. Available: https://www.nlgi.org /grease-glossary/nlgi-grade/. [Last accessed December 27, 2023].

[25] American Society for Testing Materials, "Standard Test Methods for Cone Penetration of Lubricating Grease," 2021. [Online]. Available: https://www.astm.org/d0217-21a.html. [Last accessed December 27, 2023].

[26] Polygon Composites Technology, "Case Studies," 2023. [Online]. Available: https://polygoncomposites.com /case-studies/. [Last accessed December 27, 2023].

[27] Polygon Composites Technology, "Polygon and Kuhne Industrie Delivers Composite Bearings for Demanding Mixer Cart Application," 2023. [Online]. Available: https://polygoncomposites.com/resources/ kuhne-industrie-delivers-composite-bearings-mixer-cart-application/. [Last accessed December 27, 2023].

[28] Polygon Composites Technology, "Polygon Composite Bushings Ensure Reduced Maintenance in Bulldozer Designs," 2023. [Online]. Available: https://polygoncomposites.com/resources/composite-bushings-ensure-reduced-maintenance-in-bulldozer-designs/. [Last accessed December 27, 2023].

[29] Polygon Composites Technology, "Polygon Composite Bushings Simplify Maintenance in K-Tec Heavy-Duty Earthmoving Equipment," 2023. [Online]. Available: https://polygoncomposites.com/resources /polygon-composite-bushings-simplify-maintenance-in-heavy-duty-equipment/. [Last accessed December 27, 2023].

[30] Igus Motion Plastics, "iglide® Q: Able to Withstand High Loads, Even When Subjected to the Strain of Tumbling," 2023. [Online]. Available: https://www.igus.com/info/plain-bearings-test-iglide-q. [Last accessed December 27, 2023].

[31] Franklin Fibre, "Essential Design Practices for Composite Bearings and Bushings," 2023. [Online]. Available: https:// www.franklinfibre.com /blog/essential-design-practices-for-composite-bearings-and-bushings#:~:text=For%20 best%20design%2C%20the%20length, cause%20edge%20concentration%20of%20loading.&text=% E2%80%8DThe%20smoother%20the%20surface%20of,16RMS%20or%20bet. [Last accessed December 27, 2023].

CHAPTER 4. FIELD WELDING PRACTICES

The welding of materials during production processes has become a fairly universal technique for joining metal piece parts into larger assemblies. The engineering science associated with the design and execution of welded joints is well established and highly regarded. Well-trained and experienced certified welders typically supervise and implement welding processes in manufacturing industries. However, the same degree of oversight cannot be said to take place with repairs made in the field to broken assemblies or machinery. Many repair shops utilize welders with little or no formal training. They typically lack the equipment and fixtures to properly repair a broken weld, and in general, repairs are never as strong as the original weldment. Field repairs are required in the heavy-duty trucking, agriculture, and construction industries to return broken cash-generating machinery back into operable equipment. Time is generally of the essence in these situations, and this can motivate technicians to take unwarranted shortcuts or provide substandard workmanship. It is vital that supervisors and local engineers and technicians understand the long-term problems created by this quick and dirty strategy and work to create long-lasting repairs that will survive in continued rigorous use. This review was initially published as a chapter in *Engineering Principles: Welding and Residual Stresses* and provides a general reference for making welded repairs. The review includes information on identifying specific challenges in making repairs, which weld types work best for repairs, and how to inspect repair work for quality and longevity.

ENGINEERING CHALLENGES ASSOCIATED WITH WELDING FIELD REPAIRS

TYLER J. MCPHERON, PURDUE UNIVERSITY AGRICULTURAL AND BIOLOGICAL ENGINEERING

ROBERT M. STWALLEY III, PURDUE UNIVERSITY AGRICULTURAL AND BIOLOGICAL ENGINEERING

ABSTRACT

Welding as a technology exists in two worlds. Manufacturers execute designs typically based on professional society-backed standards. Repair service centers that administer field repairs where welding applications are required can sometimes have staff members with little formal education. The challenges of a technical manager seeking welded field repairs to equipment are significant and numerous. This chapter will seek to outline the process of executing a successful welding field repair by breaking down the analysis into three parts: (1) the identification of the engineering challenges associated with a specific job, including significant stresses, difficult materials or locations, and adequate piece preparation to ensure of weld integrity; (2) the ability to properly specify the type of repair, including knowledge of the types of weld junctions and preparations, the various types of welding processes and their features, weld types and associated drawing symbols, and the repair design and repair support process; and (3) the challenges for field engineers and technical managers in identifying weld defects, executing measures, and providing adequate examination and evaluation of weld quality in the field. This chapter tries to bridge the gap between the formal, engineered welds used in manufacturing and the sometimes needed expediency of fieldwork.

KEYWORDS: welding, repairs, MIG, TIG, arc, weld defects

1. INTRODUCTION

Welded connections play a substantial role in the manufacturing processes of many types of parts, structures, equipment, and materials. Equipment manufacturers for vehicles and implements use a considerable amount of advanced welding techniques and applications throughout the design of their manufactured products. In the industrialized world, frequent breakdowns, failures, and the necessity of repairs are a part of ongoing operations. Mechanical and structural component failures are inevitable, particularly with mobile agricultural and construction equipment. Therefore, it is essential for engineers, field technicians, and those in similar roles to be educated on the characteristics of a successful weld repair, welding fundamentals, and the associated challenges of executing field repairs.

Identifying the engineering challenges of field repairs will be examined first. Properly executed welding techniques incorporate an extensive amount of engineering fundamentals to maintain the original assembly's design and ensure adequate structural integrity once complete. There are many types of welding repairs that can be categorized by location, component type, design criteria, and the degree of critical quality requirements. Understanding basic material properties and identifying the material to be repaired is important for the welding process selection and the structural analysis of the proposed repair design. Certain repairs may not be accessible enough to clean the mill scale, corrosion, or debris away, thereby narrowing the choice of possible welding processes, so repair preparation may also influence the welding process selection.

Understanding the various types of welding processes is vital to a manager seeking repairs. Each welding process features unique characteristics, often making one process more suited for a specific repair than other potential processes, based on technical characteristics of the welding process equipment. Identifying the base material properties and measuring the thickness of pieces are both easily quantifiable metrics that are used to determine the proper weld size. For economic reasons this analysis is always performed in manufacturing applications, but it is also necessary for quantifying the strength of a repair weldment to ensure that enough weld material has been applied to a given joint. Weldment strength mathematical field calculations will be examined in the chapter and illustrated with appropriate figures.

Finally, challenges for field engineers and technical managers will be summarized with emphasis on the identification of weld defects, preventative methods, and suggestions that can be used to minimize the occurrence of weld defects when conducting field repairs. Technical procedures for weld quality examination in field repairs should be useful for managers when instructing repairmen on alleviating defects or challenges in a job. Finally, the chapter concludes with a discussion of the importance of repairs to business enterprises. There is value in having basic welding technical knowledge for an array of industry specialists. This chapter seeks to promote further education and interest about the engineering properties of welding repairs and their application to practical problems and provide some value to operational managers utilizing heavy field equipment.

A variety of primary sources have been used for this chapter. Various sections of Blodgett's *Design of Welded Structures* (1972) are used heavily throughout the chapter [1]. This classic textbook incorporates a significant amount of published welding calculations, and its primary focus is on welding in industrial and manufacturing environments. However, many of the calculations and formulas are valid across all fields of welding technology and application. Information regarding material properties and analysis comes from the American Society of Mechanical Engineers (ASME), the American Institute of Steel Construction (AISC), and the American Society for Testing Materials (ASTM) publications. These organizations are responsible for writing many of the technical standards and recommended protocols for design, material testing, and property specifications in industries. Codes and practical definitions regarding welding standards are from the American Welding Society's (AWS) "Structural Welding Code" (2020) [2]. The AWS provides detailed information in its standards for a diverse array of welding applications. Various standards include code requirements, measurements, strength, specific practice recommendations, given constants for equations, education, safety practices, and other facets of welding. Where appropriate, the reader will be directed to primary sources for specific information.

2. IDENTIFICATION OF ENGINEERING CHALLENGES IN REPAIRS

There are a multitude of jobs and types of repairs requiring welding applications. Challenges can often be categorized by the location, component type, design criteria, and degree of critical quality needed in a repair. Repair location presents its own challenges from a logistical sense. Outdoor welding applications require mobilized welding equipment to access the repair. Mobilized welding equipment can require a fuel-based power source, due to remote locations or conditions. Due to the space constraint of certain repairs, it is common for mobile welding equipment to feature lengthy leads, which can also create unique circumstances. Outdoor welding repairs can present additional challenges derived from environmental factors, such as weather, terrain, and less than ideal base material circumstances. Not all welding processes are well suited for this type of environment, limiting available repair options. These challenges call for strategic planning and an understanding of certain technical properties for repairs made in outdoor environments. There are fewer constraints for indoor welding repairs, where a variety of processes are available. Indoor welding environments found in machine shops, manufacturing facilities, prototype shops, and fabrication workshops are well equipped for operator comfort and process quality without exposure to adverse environmental conditions.

2.1 REPAIR PREPARATIONS

Pipeline and structural welding are common examples of strictly outdoor welding applications. Professional welders of this category are highly skilled due to the criticalness of their weldments and their ability to produce quality welds in difficult or uncomfortable body positions. This may consist of laying under a pipe in the mud to bevel, grind, mate, and weld pipe together, or it may mean being suspended hundreds of feet in the air while welding steel beams for structural applications. Critical preparation techniques are always used during the preliminary repair preparation process to ensure the quality of the weldment, because in many circumstances time in the repair zone or position is limited.

Metal cleanliness and preparation are the most critical features for any field repair. Insufficient attention to base metal preparation will lead to unwanted imperfections in the repair. While certain welding processes are capable of penetrating through some surface rust, failing to clean the intended weld area of contaminants will always yield certain weld defects, such as slag inclusions, porosity, and craters. Weld defects lead to unwanted future cracks in the weldment, causing the component to lose structural integrity and have progressively weaker joint strength. Metal preparation is usually conducted using an electric angle grinder with a variety of disc attachments. Standard grinding discs are used to remove material quickly and perform well for removing large amounts of surface rust, paint, and other contaminates. Sanding disks, also referred to as "flap disks," are effective at removing mill scale on base material and polishing the metal surface.

Material fitment to maintain proper dimensions and alignment with other components during the welding process is nearly as vital as material preparation. Repairs due to a defect or flaw causing a component or related component to be displaced from their original state must be approached with caution. These conditions require the fitment of material to be returned to a near original state prior to the repair process. However, the removal of old material often includes the removal of the original weldment or the removal of material near the damaged area. Taking measurements beforehand or using a secondary mirrored part for dimensioning can be helpful for reestablishing the original location of a displaced component in need of repair. When proper material fit-up is completed, weld material can be added to the removed areas by making a series of weld passes to build up the filler material. Adding an amount of weld material more than what is necessary is generally recommended so that later grinding can remove the excess.

After achieving proper fitment of the base pieces, constructing a jig or welding additional structures to the member may be necessary to maintain the position of the pieces while welding. Welding induces rapid temperature changes with heat concentrated in a small area. This thermal transfer of energy causes the metal to expand slightly, and under extreme conditions this can cause warpage and deformation. High temperatures cause a crystalline structural rearrangement and reduction in tensile and yield strength in most metal materials [3]. Metal warpage is more likely to occur when applying a large amount of weld material to thinner materials with thicknesses of less than 30 mm. Part of the preparation process for the repairman is evaluating and mapping out the intended weldments in an effort to evenly distribute heat applied to the material during the repair. If material movement cannot be avoided by distributing weld material and heat evenly, a temporary fixed member can be welded in place to prevent movement from occurring. The temporary support can later be removed and ground away.

2.2 MATERIAL ANALYSIS

Welding repairs may be necessary on a variety of metal components that can be made from ferrous materials, such as carbon steels, stainless steel, and cast iron, as well as non-ferrous materials, such as aluminum and titanium alloys. Identification of the material is essential to executing a welding repair. Repairmen need to accurately be able to access which materials they are dealing with. All raw materials have specific properties associated with the type of material. Material properties are defined as measurable, quantifiable properties associated with the material. The material properties help categorize different material and ease the process of material selection. Evaluating categories of mechanical properties of material is an effective way to identify material for a field repair. Documentation is always best but seldom exists in the field. When confronted with an unknown material, investigation should include at minimum a chemical test and a hardness test. These two property indicators will help qualify the weldability of the material. Other properties can help narrow uncertainty in the base material. There are generally considered to be five categories of mechanical properties for common building materials [4]:

1. physical properties
2. mechanical properties
3. thermal properties
4. electrical properties
5. chemical properties

Physical properties are perhaps the most easily identifiable material characteristics when conducting field repairs. These properties include the shape, size, color, texture, finish, porosity, and luster of the subject material [4]. Technological properties are also referred to as basic mechanical properties for the metal, and these include hardness, malleability, machinability, weldability, and formability. It is recommended that the material's physical properties be evaluated first. This will allow an easier understanding and identification of mechanical properties. In a repair situation, material identification is important for understanding the behavior of the component's base metal and how it is likely to react to different welding processes.

Mechanical properties are critical to understanding structural repair applications, particularly when there are critical zones of stress and strength maintenance requirements. Material properties commonly found in engineering material references are the following [3]:

1. ultimate tensile strength
2. elongation
3. modulus of elasticity
4. compressive strength
5. shear strength
6. fatigue strength

Tensile strength for different types of material is experimentally determined by a standard testing method, conducted using a tensile test machine. The selected material is marked at two locations, 50.8 mm apart. Once the selected material is placed on the machine, an axial load is applied by pulling the material in opposite directions at a constant rate. As the test progresses, the load divided by the original cross-sectional area of the material within the marked area represents the resistance that the material has to the tensile load being applied [1].

The stress (σ) unit is in force per area, while the strain (ϵ) unit is formally dimensionless and is expressed as displacement in length per original length. The maximum load applied before failure of the material, divided by the static cross-sectional area of the material being tested, is equal to the ultimate tensile strength (σ_u) of material. From stress and strain values, the modulus of elasticity of a material can be calculated as [1]

$$Modulus\ of\ Elasticity\ (E) = \frac{Stress\ (\sigma)}{Srrain\ (\epsilon)} \tag{1}$$

Modulus of Elasticity (E) is a way to quantify the springiness of a material, or the stress value of a given material as it is deformed by a force in one direction. Modulus of Elasticity is also commonly referred to as Young's Modulus, after English physicist Thomas Young. The American Institute of Steel Construction states that the standard for all low-carbon steel is a modulus of elasticity of 200,000 mPa [5]. Section area is an important metric, used when calculating stress and strength of materials with loads applied in compression, tension, and shear configurations. If the member is not symmetric throughout the length of the applied load, then the section at which the material or structure will induce the most stress is used in the calculation. Once the desired cross-section is found, the neutral axis must be located. The neutral axis of a section represents the plane of zero strain and zero stress and can be a good place to locate a spot weld during the fitment process [1].

Material hardness is a well-correlated property with many other physical properties and is determined using a Brinell hardness test. The test is conducted by applying a known load to the surface of the material using a hardened steel ball. The diameter of the impression that the ball leaves on the tested material is the measured result of the test. The diameter of the impression can be converted to the Brinell number as follows [6]:

$$BHN = \frac{2P}{(\pi D(D - (D^2 - d^2)^{0.50}))} \tag{2}$$

where BHN = Brinell Hardness Number, P = Load on indenting tool (kg), D = Diameter of hardened steel ball (mm), and d = measured diameter at the rim of the impression (mm).

Fortunately, the Brinell Hardness Number does not typically need to be calculated. For most materials, the number can be found using various Brinell charts. One might be exposed to materials with a high Brinell hardness utilized in high wear environments, as in abrasive situations due to contact with other moving components. The Brinell hardness can be increased using a thermodynamic hardening process. There are various methods of hardening materials, but in the simplest form, hardening is achieved by increasing the temperature of the material to a modest degree and then rapidly cooling it by quenching the material. The quick change from a high temperature to a cold temperature hardens the material by locking in elevated temperature crystal structures.

Tool steels used for drill bits, mill cutters, and hand tools are typically hardened, along with other critical mechanical components, such as shafts, bearings, and gears. Hardened materials can be a challenge for welding repairs due to the impenetrable nature of the material. The heat applied to the material during the welding process, along with the rapid cooling typically present in welding, can make the hardened steel base material brittle and cause cracking along the joint. Heating the material slowly and evenly with an oxygen acetylene torch while monitoring the temperature of the joint before welding will soften the material and allow the weld process to penetrate deeper. After the weld is complete, slowly cooling the material around the joint will maintain the material's hardness but make the material less brittle. This softening process is also referred to as annealing [7].

The ASTM is an organization established to produce standards for material properties of all sorts. For nearly 120 years, the ASTM has written technical standards for materials, products, and other systems [8]. Physical properties are defined by characteristics, such as corrosive resistance, hardness, density, and thermal conductivity, to name a few. When choosing metal material or evaluating an existing component for repair, the corrosive resistance is an important factor to consider. Understanding the environment that the material is exposed to aids the welder in selecting preparation requirements and the welding process. In general, the welder should determine the following material properties before starting a repair:

1. the material hardness
2. the moisture exposure
3. will paint need to be removed
4. whether it is an outside or inside application

It is commonly known that mild steel is corrosive, but when painted the structural life of the steel is lengthened significantly. Corrosion rate is simply measured by the millimeters of corrosive penetration into the material per year. A common alternative to painting is plating of the metal. Materials can be plated with a variety of different plating materials. Zinc-plated bolts are a common example of a component that receives plating for increased durability. It might be best to use stainless steel for structures exposed to saltwater or high-moisture atmospheric environments. Stainless steel is highly resistant to oxidation and is a popular choice to use for marine structures, along with many industrial food-processing applications having health code precautions. Materials with any of these anticorrosion features must be treated specially, and repair welders must take these elements into account when planning repairs. Additionally, a repair technician should almost always consult a specialist if they spot a highly stressed area, find corrosion or hydrogen embrittlement, or have a high-temperature operational environment [9]. Finally, if chemical information about the base material is available, levels of carbon and other alloying materials provide critical information to the repair technical. Certain stainless steels cannot be effectively welded, and in general a 0.35% carbon level is typically considered the upper limit for welding. Material chemistry mainly identifies what cannot be welded or where extreme caution should be applied when making repairs. There are now test kits available that can determine a metal's chemical content within a matter of hours [10]. There is simply no excuse for not knowing what material you are dealing with anymore.

3. WELDING FUNDAMENTALS AND EQUIPMENT

There are three primary welding processes: shielded metal arc welding (SMAW/Stick/Arc), gas metal arc welding (GMAW/MIG), and gas tungsten arc welding (GTAW/TIG). This section will describe the characteristics and advantages of each process. Various welding positions, the different weld joint types, weldment preparations to the base materials, the calculation of weld fillet size, and weld strength will additionally be reviewed.

3.1 ARC WELDING

Shielded metal arc welding (SMAW/Stick/Arc) is chosen for a variety of applications due to its uniquely robust properties, allowing it to be used in many environments. Stick welding is heavily used in structural and industrial heavy metal applications when less than ideal conditions are present. Common SMAW/Stick/Arc welding joints include the assembly of structural frameworks for buildings and the joining of pipe segments together in pipeline applications. The connections of high-pressure pipeline segments are generally made by SMAW/Stick/Arc welding. The pipe ends are typically beveled, mated, and welded with an E6010 electrode. This is called a root pass. There can be several root passes depending on the application. Root passes are then overlaid and capped by several additional weld passes, building additional filler material with each pass. This kind of weld is durable and self-healing around small weld defects. It is the oldest form of electrode welding and still preferred when deep penetration is required.

There is an assortment of consumable electrodes available to the repairman. These are also referred to as "rods." Electrodes are sticks of filler metal 30 *cm* long wrapped with a flux coating. As shown in Figure 4.1, they are referenced with a four-digit numerical code, with each number referencing a specification for the

FIGURE 4.1. Electrode specification stamped on the flux coating of a typical welding rod.

FIGURE 4.2. Stick welded components showing the splatter form the process.

electrode. The first two numbers reference the tensile strength of the weld material that the rod will produce. The third number refers to the position(s) the electrode can be used for welding. A 1 means that a rod can be used in all positions, and a 2 means that it can only be used in the flat and horizontal positions. The fourth number refers to the type of flux coating, the hydrogen content of the electrode, and if the rod can be used with alternating current, direct current, or both. An advantage of the stick welding process is its ability to make high-penetrating welds on imperfectly prepared material with oxidation and mill scale, the oxidized scaly substance on surface of most hot rolled material.

This makes SMAW/Stick/Arc welding an excellent choice for field repairs. The stick welding process is commonly used by repair shops for agricultural, off-road, and construction equipment repairs. With the correct rod, rusty and dirty metal can be welded effectively with minimal preparation. This process produces a very strong weldment with high material penetration and can produce an aesthetically pleasing weld bead using certain electrodes. Unfortunately, as shown in Figure 4.2, stick welding generates far more splatter and slag peel than other welding processes. However, those other processes require clean metal with minimal mill scale to make solid weldments. Stick welding can operate with either direct current or alternating current. Stick welding machines use a ground clamp and an electrode holder, commonly referred to as a "stinger," to make the necessary electric circuit.

SMAW/Stick/Arc welding is the generally preferred choice of process by many repairmen for field repairs. As previously mentioned, field repairs often require the use of mobile welding equipment. Notable welding equipment manufactures such as Lincoln Electric and Miller Electric manufacture gasoline- and diesel-powered welders for this purpose. Test engineers, pipe fitters, construction crews, and other field repairmen will generally have their service vehicles equipped with machines of this category. Mobile welding rigs typically have electrode holders and ground leads in lengths of 15 to 30 m for repairs and jobs, where only limited vehicle

FIGURE 4.3. Spool of 0.58 mm filler wire mounted on a MIG welding machine.

FIGURE 4.4. Horizontal fillet weld on clean material done by GMAW/MIG welding.

access is available. If the stick welding process is utilized for repair work to an appearance-critical piece of equipment, extreme care must be taken to effectively tarp and shield the balance of the unit from the inevitable splatter of the process.

3.2 MIG WELDING

Gas metal arc welding (GMAW/MIG) is the predominate joining method in manufacturing and industrial settings. This process is preferred because of its weld quality and speed, making it less expensive to implement in production situations. GMAW/MIG welding produces very little slag and creates a very clean welded connection when welding clean material. MIG welding is often referred to as "wire feed welding." As shown in Figure 4.3, its nickname is due to the spool of wire fed through the MIG gun by the machine. When the trigger on the gun is actuated by the operator, electrified wire is fed through the end and shrouded by a gas cloud. A typical combination of shielding gas for welding mild steel is 75% argon and 25% CO_2. As the welding wire is pushed toward the base material, it strikes an electric arc and melts the target material and filler materials together. Filler metals are rated by their tensile strength. A tensile strength rating of 482 MPa is common for most spools of filler wire. MIG welding is not as effective as SMAW/Stick/Arc for use on oxidized material without significant cleanup and material preparation. However, it is more time-efficient than other processes for clean assemblies, making it an excellent process in production environments. As shown in Figure 4.4, it makes a beautiful, clean weld on clean material. Materials such as steel, aluminum, magnesium, carbon steel, nickel, and other alloy metals can be welded with the GMAW/MIG process, making this a versatile industrial tool when multiple materials are in use. The ability to weld a variety of materials makes GMAW/MIG a top choice in high-production manufacturing [11].

3.3 TIG WELDING

Gas tungsten arc welding (GTAW/TIG) is a process commonly used for specialty welding repairs and applications, where additional craftsmanship is necessary. The GTAW/TIG welding process is commonly used on components that are made of stainless steel and other non-ferrous metals. TIG welding is expensive and time-consuming due to its precision, but as shown in Figure 4.5, the quality of the GTAW/TIG weld is unsurpassed. Consequently, it is not often used for facilitating field repairs but may be used for smaller repairs, especially for stainless steel repairs. Stainless steel is an alloy with unique properties that do not allow it to be welded easily with other welding processes or without specific equipment configurations. Stainless steel material is subject to cracking at high welding temperatures from other welding processes, causing additional weld defects, which result in weld-quality issues. However, stainless steel can be welded nicely with the use of a TIG welder due to its low heat application from the tungsten rod in tandem with an argon shielding gas. The user has the ability to control the welding heat with great precision. TIG welding uses a foot pedal to control the machine amperage. It can operate on direct current or alternating current, but the tungsten electrode rod is not actually consumed very rapidly. The process requires filler metal to be fed manually by the welder into the arc that is struck by the tungsten rod attached to what is referred to as a "TIG torch." TIG welding is commonly used in applications where bead aesthetics are preferred because it produces no weld slag. These applications include stainless steel exhaust headers, pipe/tubing, and roll cages. TIG welding produces a high tensile strength weld and is a very good choice for a variety of applications requiring clean

FIGURE 4.5. GTAW/TIG fillet weld on aluminum material.

finished beads or the repair of exotic materials. It is a well-suited process for applications requiring clean, good-looking welds and is a highly regarded choice for critical stainless steel joints and connections.

3.4 WELDING POSITIONS

The AWS identifies four basic welding positions. These positions characterize the position of the electrode in relationship to the object being welded. The four welding positions are described as the horizontal position, the flat position, the vertical position, and the overhead position. These welding position definitions have been established by the AWS due to the subtly but critically differing techniques associated with each position. Professional welders must pass welding application tests in each position to prove their competence for all of the established welding positions [3]. Weld metal behaves differently in all positions, creating challenges to consider when conducting a field repair. Identifying where the welding material will travel due to gravity is a major consideration when making repairs in each welding position. Table 4.1 summarizes several of the welding symbols associated with different welding positions and type of weld. Well-specified repair instructions identify the welding position, weld type and joint preparation to be used.

3.4.1 FLAT POSITION

The flat position is the simplest position and the easiest to execute by novice welders. Most welders find welding in this position to be natural and comfortable, typically traveling from left to right. In fact, it is generally

TABLE 4.1. *Welding positions and weld types identified by welding symbols for technical drawings*

WELDING POSITIONS AND SYMBOLS		
POSITION	**WELD TYPE**	**SYMBOL**
Flat	Fillet	1F
Flat	Groove	1G
Horizontal	Fillet	2F
Horizontal	Groove	2G
Vertical	Fillet	3F
Vertical	Groove	3G
Overhead	Fillet	4F
Overhead	Groove	4G

recommended that the component being welded be turned to the flat position if possible for easier constructability. Figure 4.6 shows the location of the material and fillet in a flat position weld. Molten weld material travels downward naturally in the flat position, penetrating the material adequately. The flat welding position is executed similarly for both fillet and groove joint configurations [12].

3.4.2 HORIZONTAL POSITION

The horizontal welding position is similar to the flat position, with the joint rotated at 90 degrees. A fillet weldment in the horizontal position can be slightly more challenging to execute than weldments made in the flat position. Horizontal groove welds typically require more attention and control of the weld pool to establish an equal penetration into both base metals. Again, left to right travel is generally preferred. Figure 4.7 shows the location of the material and the fillet in a horizontal position weld.

3.4.3 VERTICAL POSITION

The vertical position requires traveling uphill or downhill when welding. The vertical welding position is illustrated in Figure 4.8. It can be difficult to perform due to the downward gravitational pull on the weld material before it cools. This can cause the weld material to build up when traveling uphill and can extinguish the arc when traveling downhill. To decrease the occurrence of material buildup, welders typically lower the amperage slightly on the machine. This allows the weld material to cool faster, decreasing its ability to move gravitationally. Traveling uphill is the preferred direction of travel for vertical position welding. This gives the welder more weld pool visibility without the weld material blocking the welder's view of the arc. The weld pool will cool away from the arc, so the proper speed of travel is critical to executing a proper vertical weld.

3.4.4 OVERHEAD POSITION

The overhead welding position is the most difficult position, even for the most skilled tradesman. Figure 4.9 illustrates some of the challenges posed by the gravitational vector in executing an overhead weld. In the overhead position, the weld pool wants to naturally flow in the opposite direction of the joint. This will cause weld material to sag away from the joint as it cools, creating a crown [12]. A crown is excess concavity of the face of the weld material that sags away from the joint [13]. Poor overhead welds can have void spaces, so keeping the

FIGURE 4.6. Base material orientation and working position for a flat position fillet weld (1F).

FIGURE 4.7. Base material orientation and working location for a horizontal position fillet weld (2F).

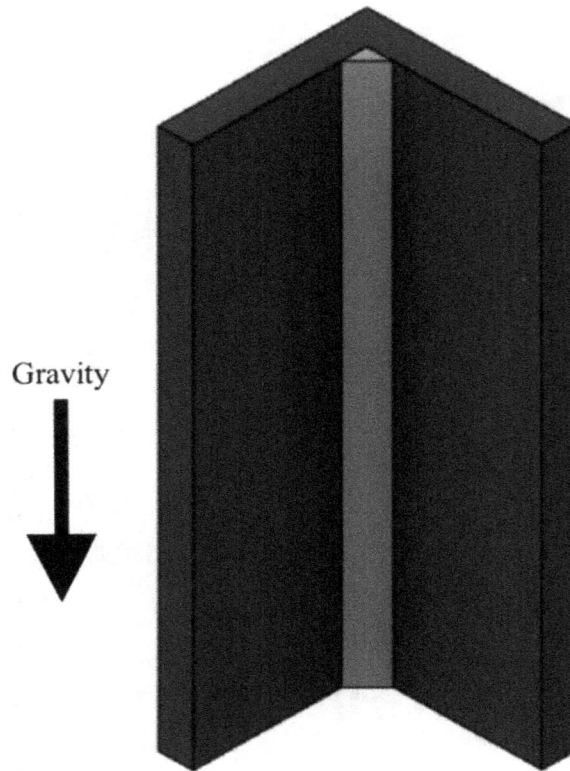

FIGURE 4.8. Base material orientation and working location for vertical position fillet weld (3F).

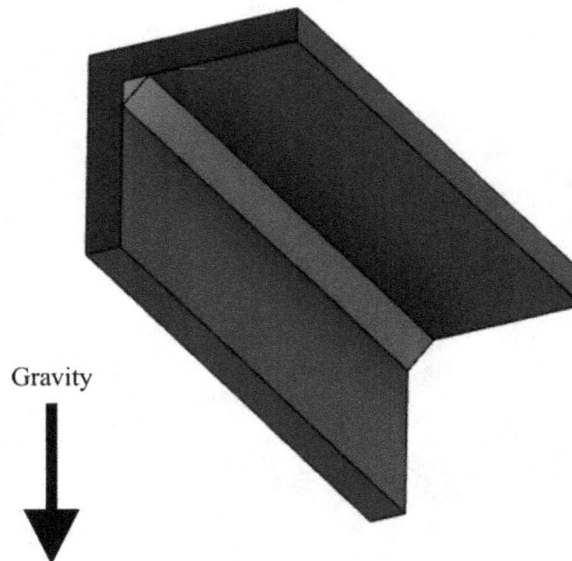

FIGURE 4.9. Base material orientation and working position for over-head position fillet weld (4F).

arc tight and the diameter of the weld pool small helps avoid this defect. Welding in the overhead position is generally avoided if possible, but if required, additional care must be taken to identify problems and correct deficiencies. A project manager must understand that this weld is extremely challenging and should be approached with caution in a repair.

3.5 WELD JOINT TYPES

In a welding repair scenario, there will have been a joint already constructed. Identifying the joint type is critical so as to know what type of weld should be applied to repair the member. There are five primary joint configurations established by the AWS, shown in Figure 4.10. The five weld joint configurations are described by the physical intersection between two joining materials. The shape of the base pieces is irrelevant to the joint classification. All metal joints can be classified under one of these five categories, whether the joint is constructed on plate, pipe, or premade structural shapes. Weld joint classification is different from weld type. The former depends on the relative position of the base pieces to one another, while the latter depends primarily on the joint preparation. In general, there are three weld types: fillet, groove, and slot. The various weld types are illustrated in Figure 4.11.

3.5.1 FILLET WELDS

The fillet weld is a type of weld used in most applications and is notable for being the most common type of welded joint connections. Fillet welds are used to make lap joints, corner joints, and T-joints [14]. The basic shape of a fillet weld is a right triangle with two equal leg sizes. However, it is difficult to make a fillet weld that formulates a perfect right triangle. There is a reasonably generous tolerance for the shape of a fillet weld. The weld throat is assumed to be the length of the hypotenuse across the weld shape. Fillet welds must maintain a certain size relative to the thickness of the base metal. In other words, the throat of the weld cannot be too convex or too concave. A fillet weld with a throat that is too convex means there is not enough penetration into the joining material. A fillet weld with a throat that is too concave results from not enough weld metal applied. Either condition creates a weak weldment, and both should be flagged for repair in the present.

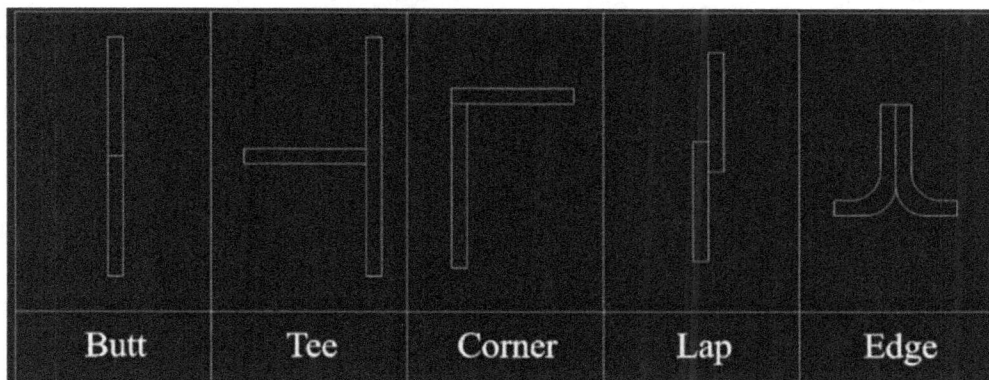

| Butt | Tee | Corner | Lap | Edge |

FIGURE 4.10. The five established weld joint configurations.

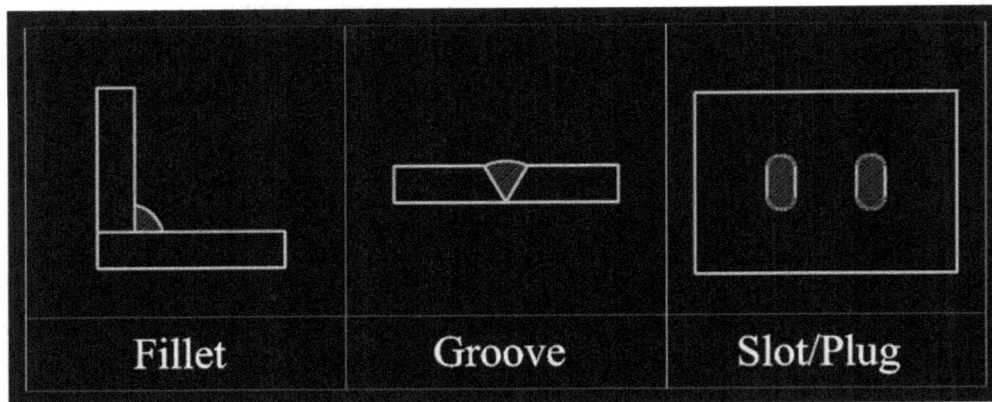

FIGURE 4.11. The three established weld types.

3.5.2 GROOVE WELDS

Groove welds are classified by the various shapes and sizes of the preparation on the base materials. The types of groove weld classifications are shown in Figure 4.12 and listed below:

1. square groove
2. V-groove
3. bevel groove
4. U-groove
5. J-groove
6. flare V-groove
7. flare bevel groove

Groove welds can be used in corner joints, T-joints, and butt joints and are often used for tight fitment applications. Each of the groove weld types listed above has its own symbol and associated dimensioning method per the AWS symbol standard. There are significant material preparation techniques associated with groove weldments, and they typically concern the construction of the specific groove. These details are specified in the weld symbol structure. Beveling the ends of the material in a specific manner is necessary to form the groove. The beveling extent is specified by different bevel angles. Bevels allow for weld metal to fill the beveled-out area more easily and increase the strength of the weld.

Figure 4.13 represents the material preparation of two steel plates for a 60° V-groove weldment. The angular dimension for the groove is specified in the weld symbol structure. After the weld has been applied between the pieces of base metal, the excess can be ground away to present a smooth surface of the steel plates, as if the plate is one piece. If executed properly, groove welds can provide superior penetration and strength.

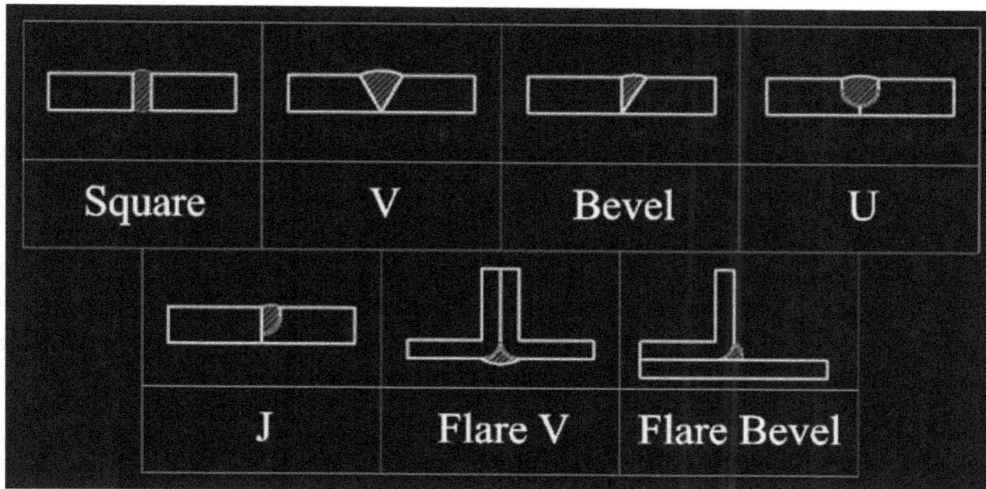

FIGURE 4.12. Groove weld preparations in base material to be joined.

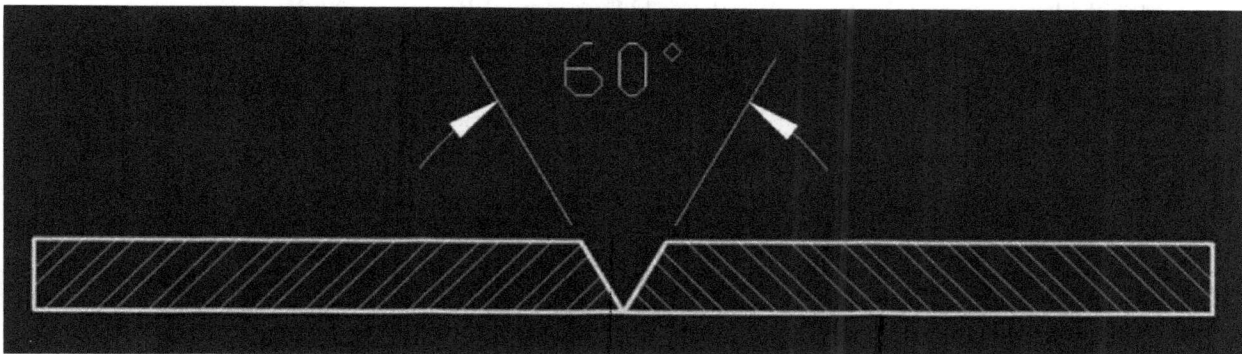

FIGURE 4.13. Material preparation for V-groove weldment.

3.5.3 PLUG AND SLOT WELDS

Plug welds and slot welds are similar and are often used to join two overlapping plates. One of the two plates will have drilled holes or milled-out elongated slots. This internal opening to the second piece is where the weldment will be produced. When dimensioning a plug or slot weld, the hole diameter will be shown to the left of the plug symbol, and the plug spacing (distance between each hole) is shown to the right of the symbol. For slot welds, the width of the slot is shown to the left of the symbol, while the length and pitch are shown to the right of the symbol. A drawing detail is often referenced in the tail of the weld symbol structure. If the slot or plug is not to be filled completely with weld metal, the fill depth will be specified [14].

3.6 DETERMINING WELD SIZE

Determining the proper size of a weld is essential in conducting a welding repair. Overwelding by repairmen does not generally enhance the chances of success for a repair. A common misconception is that applying an excess of weld material provides additional strength to the joint. This is false, and overwelding can actually weaken the joint by the additional application of heat transferred to the material around the joint. Overwelding is quantifiable and can be calculated in manufacturing applications to minimize production costs. For repairs, overwelding adds time to the job without any benefit to the customer and may be counterproductive. The depth of weld penetration is a more important factor in creating a good weld and can be properly gauged by correctly sizing the thickness of the weld.

The thickness of the weld is controlled by the depth of material being welded. It is critical to understand that the material thickness input must be the thinnest material associated with the joint that is being welded. This is especially important in weldment strength calculations. *Design of Welded Structures* defines the recommended size of a fillet weld using Equations 3 and 4 [1]:

$$L_S = \frac{3}{4} * (M_t)$$ (3)

$$W_t = cos(45°)(L_S)$$ (4)

where M_t is the base material thickness, L_s is the weld leg size, and W_t is the weld throat size.

Figure 4.14 illustrates these variables in an illustration of a cross-section of a tee joint. Following this recommendation for the fillet size will create the proper amount of weld penetration into the base material for a solid joint.

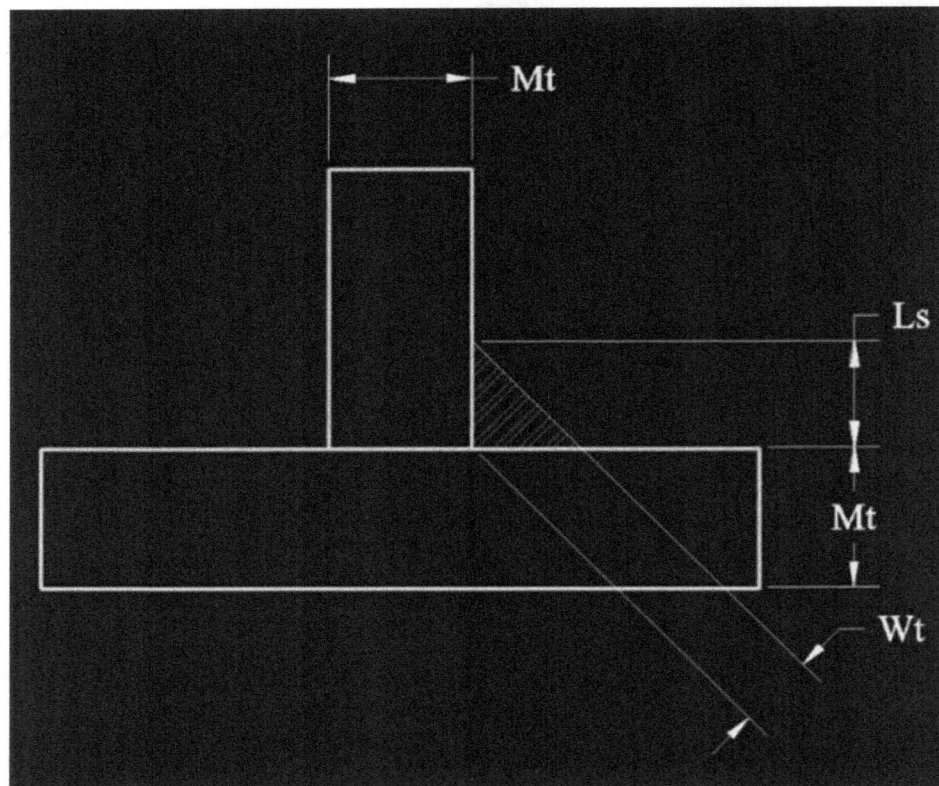

FIGURE 4.14. An illustration of fillet weld size variables used in weld strength calculations.

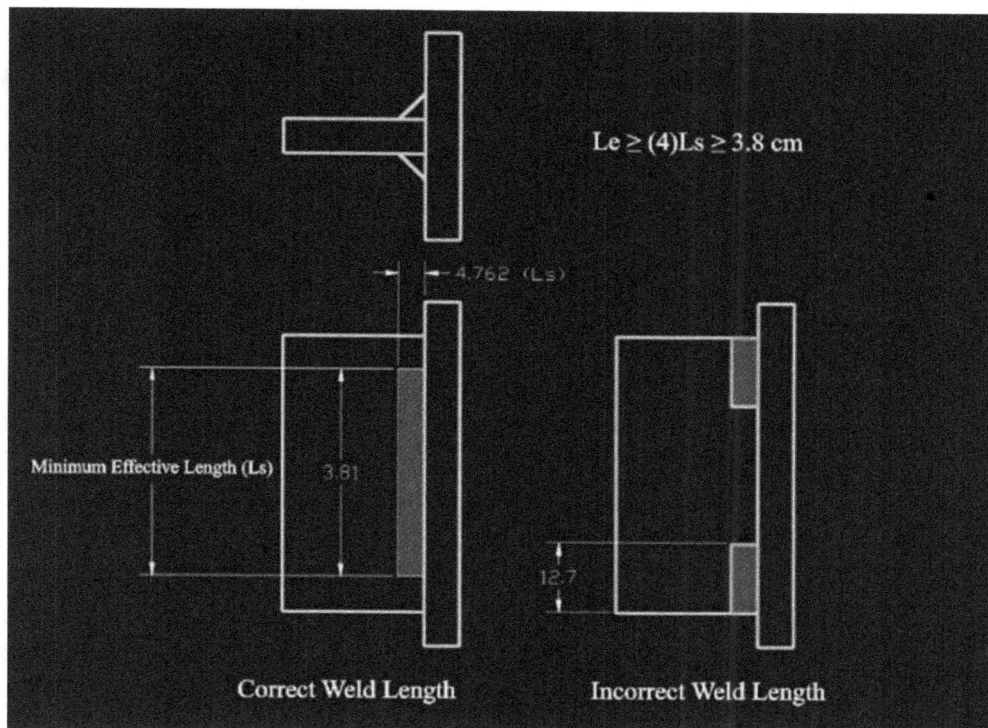

FIGURE 4.15. Minimum effective weld length of run.

When calculating the throat size, a 0.71 scale factor sometimes appears in the equation for calculating throat size. This is the rough equivalent of cos(45°). Depending on the structural application, type of stress on the effective weldment, and how much stress the weld will endure, different welds will require different lengths of run. Determining the proper length for a weld minimizes the potential for overwelding. For instance, if two plates needed to be joined but no load or minimal load is applied in shear, then the most effective way to join the two plates would be applying a series of small stitch welds across the connection. Stitch welds are intermittent rather than continuous welds. Fillet welds used to transmit forces should not be less than four times its leg size in length or 3.8 *cm*, whichever is greater [1]. Neither should the effective weld size of the fillet weld exceed one-fourth of the length of the weldment [1]. Figure 4.15 presents these fillet length constraints graphically.

Stitch welding is a technique used to connect base material pieces that will see little stress and fatigue in their duty cycles. Additional weld fillet length beyond the structural need is wasteful and costly. Stitch welding is generally a great technique for field repairs that require lengthy connections, eliminating the need to weld a solid bead that would introduce a significant amount of heat into the base material, potentially causing distortion and warpage. Engineering drawings feature unique callouts for stitch welding that include dimensions referencing the length of the stitch weld, the distance between each stitch weld, the weld type, and the size of the weld [15]. The drawings will also feature callouts for the intermittence of the weldment, such as intermittent welds on one or both sides of the joint and the staggered intermittent interval for the weldments [15]. In field repairs, it is usually up to the repairman to evaluate the stitch weld dimensions and the weld run intermittence, depending on the stress of the joint. Significant guidance for the design of the weld can be had through the proper estimation of the needed weld strength.

3.7 WELD STRENGTH

Weld metal has tremendous strength, generally higher than the raw strength of the base metals that are being joined. This strength can be calculated in relation to the type of load the material will face. Welded components may endure multiple different types of stress in unison. Fortunately, the superposition of loads principle allows the calculations of each type of stress to remain separate. Transverse and shear loads represent the two most common load orientations.

The fillet welds represented by the right triangles shown in Figure 4.16 are very resistant to a transverse type of loading. However, if the load follows the direction of the arrow and only one fillet weld existed on the right-hand side of the plate, the connection would be weakened and would likely bend over itself and break off at the joint, toward the right-hand direction. In general, the best practice would be to weld both sides. However, if the application required only one fillet weld due to a low load application, then the weld should be placed on the side of the plate that places the weld in tension, not compression. In this case, it would be the arrow side of the plate. If the load was in the opposite direction, then the preferred weld location would move to the other side.

Transverse loading is only used when the load applied is perpendicular to the weldment. If the load is present in a direction parallel to the weld, as shown in Figure 4.17, then the strength of the weld must be calculated in shear. Shear loading is more detrimental to the structural integrity of a weldment than a transverse load.

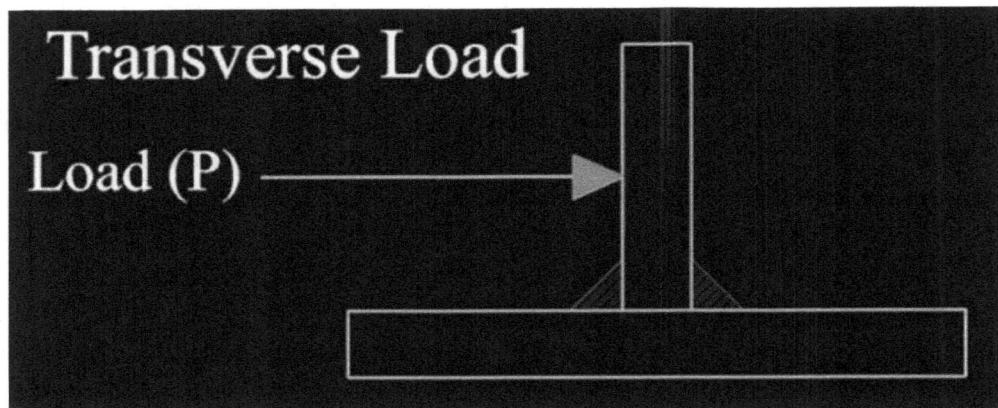

FIGURE 4.16. Illustration of transverse loading on weldment.

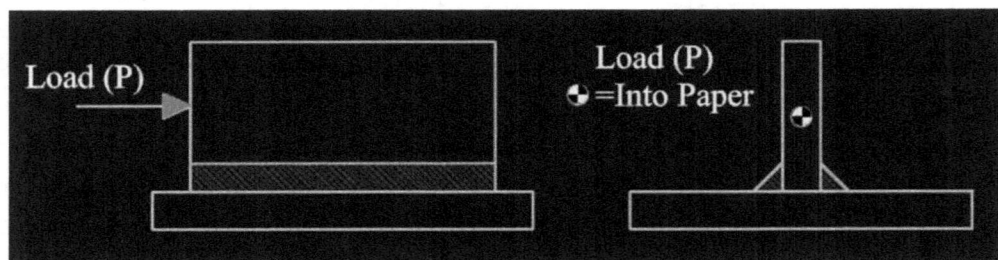

FIGURE 4.17. Illustration of shear loading on weldment.

The decreased structural strength of the weld means that the weld strength cannot be calculated using the full tensile strength of the filler metal. Instead, the AWS recommends that the tensile strength of the filler material be reduced by multiplying by a value of 0.30, reducing the tensile strength by 70% [2]. Otherwise, the strength calculation process is the same.

The calculation of weld strength for transverse load is straightforward, but it is important to emphasize that this is the calculated static strength of the weldment without measurement against an applied load. Transverse loading requires three basic inputs for determination: filler tensile strength (T_F), thickness of base material (W_t), and length of weld applied (L_W). Equations 3 and 4, along with Equations 5 and 6, are used to successfully calculate the strength of a weld under transverse load [1]:

$$A = W_t * L_W \tag{5}$$

$$F = T_F * A \tag{6}$$

where A is the sectional area of the weld, W_t is the weld throat size, L_w is the length of the applied weld, T_F is the tensile strength of the weld filler material, and F is the resistive force the weld is capable of supporting.

Table 4.2 shows the application of this model in calculating the weld strength for material and weldment length using specific values. The tensile strength for most common MIG filler wire is 483 *MPa*, and that has

TABLE 4.2. *Spreadsheet calculation for weld strength under transverse and shear loading*

STRENGTHS OF WELDMENTS			
INPUTS			
NAME	**VALUE**	**UNIT**	**DESCRIPTION**
TF	482	*mPa*	Tensile Strength of Filler
TM	3.18	*mm*	Thickness of Material
LW	1092.2	*mm*	Length of Weld Applied
TRANSVERSE LOADING			
NAME	**VALUE**	**UNIT**	**DESCRIPTION**
LS	2.38	*mm*	Leg Size
TS	1.25	*mm*	Throat Size of Weld
A	1366.26	*cm²*	Effective Area of Weld
F	**658536**	*N*	Strength of Weld
NAME	**VALUE**	**UNIT**	**DESCRIPTION**
STF	482	*mPa*	Tensile Strength of Filler
STM	3.18	*cm*	Thickness of Material
SLW	1092.1	*cm*	Length of Weld Applies
SHEAR LOADING			
NAME	**VALUE**	**UNIT**	**DESCRIPTION**
SF	**197561**	*N*	Strength of Weld

been used here. The calculations show that the maximum strength of the weldments under transversal load is 200 *kN* and is 198 *kN* in shear load. This number would be an upper bound for loading, because the calculation lacks any accounting for imperfections. Any weld imperfections, defects, undercut, material imperfections, material fit-up problems, or dynamic or thermal loading during welding will reduce the strength of a welded joint. This calculation merely provides the theoretical strength of the specific design under specific conditions. In manufacturing, weldment specifications are properly simulated using finite element analysis, tested, and inspected to ensure that the applied loads do not exceed the factor of safety in the design. However, this level of analysis and design is typically not available to repair specialists. They simply tend to increase the factor of safety within their designs accordingly.

All of the necessary parameters for the basic weldment strength calculations can be easily obtained when conducting field repairs. A tape measure and a phone calculator are the only tools truly needed to execute this calculation in the field. The tensile strength of the filler wire can be found on the wire spool attached to the welding machine. For arc welding applications, the tensile strength is called out on the electrode. Proper weldment design in repairs is just as critical, perhaps more so, than in original manufacture.

Stitch welding is a technique used to connect base material pieces that will see little stress and fatigue in their duty cycles. Additional weld fillet length beyond the structural need is wasteful and costly. Stitch welding is generally a great technique for field repairs that require lengthy connections, eliminating the need to weld a solid bead that would introduce a significant amount of heat into the base material, potentially causing distortion and warpage. Engineering drawings feature unique callouts for stitch welding that include dimensions referencing the length of the stitch weld, the distance between each stitch weld, the weld type, and the size of the weld [15]. The drawings will also feature callouts for the intermittence of the weldment, such as intermittent welds on one or both sides of the joint and the staggered intermittent interval for the weldments [15]. In field repairs, it is usually up to the repairman to evaluate the stitch weld dimensions and the weld run intermittence, depending on the stress of the joint. Significant guidance for the design of the weld can be had through the proper estimation of the needed weld strength.

4. CHALLENGES FOR ENGINEERS AND TECHNICAL MANAGERS

One of the more difficult tasks for an equipment manager is signing off on the acceptance of repair work when the work falls within the domain of a specialist. Welding repairs certainly meet these criteria for most individuals. This section will examine weld defects and how to recognize them and will cover the repair of defective welds and then go into detail about more extensive examination methods necessary to verify the quality of a critical weld.

4.1 WELD DEFECTS

Defects in repair welds can have calamitous results. Therefore, it is necessary to identify defects and imperfections in a repair weld before returning equipment to duty. Understanding the allowable variations in weld material's physical and mechanical structure will provide a reasonable basis for inspecting welds for imperfections. Pockets of impurities and variations in the weld that are within the acceptable tolerance range are called "discontinuities." Discontinuities that exceed the acceptable tolerance are called weld "defects" [16]. Defects must be ground away and repaired. Defects are generally caused by poor welding technique, poor joint fit-up, or both. In the world of repair, poor joint fit-up is the most common cause of defective weldments. Repair jobs often consist of components that have separated and are no longer in their original position, making it difficult for the repairman to achieve the original geometry. This can be accomplished for many repairs, but sometimes it requires significant attention to detail and a bit of creativity to manipulate the original material into an acceptable repair position.

Defective welding technique can cause impurities that affect the shape and size of the weldment, cause imperfections in the internal structure of the weld, and create other defects that adversely affect the weld's strength. The basic categories of weld defects are overlapping, undercutting, distortion and warpage, cracks, craters, porosity, and inclusions. Overlap is excess weld metal that reaches beyond the joint onto the base metal. The primary cause of overlap is an incorrect angle of the electrode with respect to the base material. If the electrode is not angled away from the weld pool on the leading edge, the arc will manipulate the weld pool, causing notches to form between the overlap and the base metal.

Undercutting defects occur when electrode travel speed is too fast, causing grooves or gaps on the toe of the weldment between the base metal and the weld metal. The rapid travel speed does not allow enough time for the weld metal to fill in after penetration has occurred into the base metal. A thin or narrow weld bead also indicates excessive travel speeds. Good technique and weld speed will eliminate this problem.

The misconception that overwelding creates stronger joints was mentioned previously. Overwelding can also cause distortion and warpage of the base metal by the excessive application of heat. This occurs due to an improper joint design from the base metal thickness and dimensions. Applying a large amount of weldment in several passes can induce a large amount of heat and will also cause thin materials to flex and warp. This weld defect is extremely difficult to correct after the fact due to the alteration of the base metal's natural shape. Heavy clamping in critical locations, before welding materials are likely to deform, is usually best. Once a piece has deformed into a complex shape, corrective choices become limited and generally involve minimizing further damage.

Cracks in the weld metal are often caused by other weld defects or excess stress induced at the joint. There are several classifications of cracks. Some cracks are internal and cannot be identified from the surface or at the throat of the weld. The stress-induced area at the toe of weldment is a common origination site for weld cracks [16]. Craters can form in the weld bead, causing a gap or hole in the weld metal due to the lack of filler metal deposited. Craters are easily corrected by revisiting the defect area after the weldment is complete. A new electrode should be used, beginning the tie-in 2.5 *cm* or so in front of the crater, allowing weld metal to melt and deposit into the crater and extinguishing the arc after proper amount of weld metal has filled the depression.

Porosity occurs when gas bubbles become trapped in the weld metal. This creates an uneven distribution of foamy weld metal. Porosity weakens the weld and can generally be identified at the surface of the weldment. Visible porosity is an indication that there is also porosity located below the throat of the weld. Inadequate weld preparation and contamination are the most common cause of weld porosity. The trapped gases are usually not shield gases from the welding process but instead are gases from the oxidation and vaporization of the contaminates on the surface of the base metal [16].

Slag inclusions can result from trapped welding slag in the weld metal. This occurs only when using the SMAW/Stick/Arc welding process, since GTAW/TIG and GMAW/MIG processes produce little slag peel. Slag inclusions can happen when an electrode has been consumed and the welder ties into the old weldment with a new electrode before removing the slag peel on the surface. Defects caused by improper welding technique can be easily eliminated by following established welding procedures and maintaining proper travel speed, arc length, and electrode angle recommended for the specific welding process.

4.2 REPAIRING WELD DEFECTS

Welding defects must by repaired by the complete removal of the defect using either torch removal or grinding methods. Portable grinders with stone wheels are an excellent tool for the removal of weld defects. There is a selection of sizes in grinding wheels, and some wheels are designed to use the surface of the wheel, while others are designed to be used only on the edge. Grinding wheels designed for grinding with the outer edge of the wheel are a popular choice for removing joints where a groove or bevel is present. Once the defect has been ground away completely, another welding attempt can be tried. For long stretches of removal, an oxygen acetenyl torch may be used to gouge out the defect. This can be a more efficient method of removal in certain situations. Torch gouging may also be the only practical choice if the joint is space-constrained with little room to reach the defect area with a grinding wheel.

4.3 WELD QUALITY AND EXAMINATION

It is important for field engineers and welders to evaluate the quality of their weldments. Each kind of defect displays unique characteristics, simplifying the identification and analysis process. Some defects are visually obvious, but a surprising number of repairmen do not thoroughly inspect their work after completion. Visually inspecting the weldment is vitally important to ensure the integrity of the repair and does not require significant effort. Cleaning the weldment and the area around the joint by removing slag with a wire brush is an easy process that results in improved visibility. The visual examination of a weld can usually be conducted with a flashlight, looking closely for the undesirable characteristics associated with the defects previously discussed.

The best practice for larger repairs and welding jobs is to have the inspection process functioning throughout the repair, conducted by a trained inspector at each step of the way. This preventative inspection protocol eliminates the continuance of defects during a repair [1]. However, an independent inspector for field repairs is not usually feasible. Therefore, it is vital for the welder to be capable of identifying defects as they occur. Quality control can be improved by having two or more qualified persons provide inspections. A fillet weld gauge is a handy tool for determining the size and quality of welded field repairs. This device is a measuring tool for checking leg size and throat size of the weld [17].

In specialty applications where structural integrity is of utmost importance, inspections must be regular, and welds must meet the acceptable defect tolerance level. Specialty applications, such as underwater bridge welding, ship welding, and oil rig applications, must adhere to strict guidelines and codes and are classified as Class A welds. Welders in these fields must be licensed by passing a series of weld tests for joints configured in the 3F and 4F positions, administered, and evaluated by AWS-certified inspectors before beginning a new job, regardless of the prior experience of the welder. This requirement is due to the potential ramifications of a failed weldment [18].

4.4 FILLING GAPS AND REPAIRING CRACKED COMPONENTS

Cracked components are common in the mechanical industry and are a regular failure mode for many pieces of equipment. These repairs are some of the toughest to execute, and this section will illustrate the overall process so that the equipment manager has an idea of the complexity of the operation. To repair a crack in carbon steel, a bevel directly on the crack must be ground to allow the weld metal to penetrate the base material, creating a tight joint at the crack. An electrode diameter size close to the width of the groove made by the grinder must be used. A standard size filler of 0.89 mm wire is generally sufficient if using a MIG welding process. Depending on the size and characteristics of the crack, an E6010 electrode capped with a low-hydrogen E7018 electrode would be recommended to eliminate any undercut or hydrogen deposits created by the E6010 electrode if using a GAW/Stick/Arc process. Finally, the weldment should be ground flush with the base metal to maintain its original appearance.

A field technician might encounter a situation where a gap needs to be bridged between two base pieces. A gap is when two materials should be welded together but the pieces should not contact each other, as in Figure 4.18. The gap must be filled with weld metal, as in Figure 4.19. This technique should be considered as the second stage in repairing a crack. A break that needs to be fixed by bridging a gap often originates from a hairline fracture or crack. When the material separates, it cannot always be drawn back together for repair. Filling gaps can be done in a variety of ways with different welding processes. Large gaps can be filled by

FIGURE 4.18. Tack welds applied to establish a 0.3 *cm* gap between the separated pieces.

FIGURE 4.19. Multiple large tack welds used to fill the gap between the separated pieces, where a continuous weld would likely blow through the gap.

using a backing strip. A backing strip is a section of material tack welded or stitch welded onto the base metal. This provides a surface for multiple welding passes without burning through. If a backing strip is used, it can later be ground off and removed or can be left in place, depending on the criticality of the original dimensions. One simple technique to execute a gap fill is by making a series of tack welds along the joint, evenly spaced out. This allows the joint to be welded solid, while the tack welds provide enough weld metal to not blow through the joint. The fillet is typically ground flush after sufficient penetration and connection between the pieces has been established, as in Figure 4.20. Finally, a finish fillet can be applied to cap the weld, as in Figure 4.21.

FIGURE 4.20. Welds ground down to remove defects that occurred during the tack welding process joining the two separated pieces.

FIGURE 4.21. A fine cap fillet pass used to finish joining two separated pieces, which may be ground flush for a solid appearance.

5. CLOSURE AND VALUE OF WELDING TECHNICAL KNOWLEDGE

Field repairs are frequently needed and often encountered by many technical professionals. Certified welders in manufacturing, pipeline, industrial, and fabrication shops gain a large amount of theoretical- and application-based welding knowledge during their skilled trade education. On the other hand, engineers and technologists are typically provided with very basic knowledge of welding processes and applications and have minimal field experiences. These career areas may experience the need for welding repairs, especially when prototyping and testing. Gaining additional knowledge to further understand theoretical process and mathematics behind welding techniques will increase the ability to execute, facilitate, and manage repairs of this nature. Management professionals will likely encounter the need to retain welding services, and they need to be able to competently specify the work to be done and be able to inspect it.

Challenges are present in all welding repairs. Critical considerations of these challenges are often not made due to a lack of knowledge surrounding some of the more advanced topics. It can be difficult to evaluate all the factors in field repair environments when time is constrained. Material properties, hardness of material, and stress characteristics should be thoroughly investigated when the repair is critical. Welding processes, equipment, and weldment preparation can provide the basis for conducting a proper field repair or general welding task, and it is incumbent upon a good equipment manager to understand the basics of welding. A breakdown of fundamental welding positions and types of welds has been shown. It has been demonstrated that the repairman's physical position in relation to the weld is important to constructing a proper joint and executing a sufficient weldment. It was shown that the calculation for the correct weld size in relation to the base material is a simple field calculation. In more advanced major repairs where stress-induced members are present, further field calculations for structural analysis of the weldment can be performed without excessive computation. These calculations provide quantitative metrics to determine the integrity of the repair weldment design if it is to be applied properly. The repairman can then judge whether the proposed weldment is adequate and sufficient for the repair.

The identification of welding defects, imperfections, and the allowable tolerances for each is imperative in evaluating welds and managing quality repairs. The difference between an acceptable weld and an unacceptable weld can be minimal depending on the defect, and it is critical to be able to differentiate between the two. Finally, knowledge regarding the repairs and the correction of welding defects is perhaps the most valuable skill an equipment manager can have, as the correction of these deficiencies is frequently among the most common of needed field repairs.

6. ACKNOWLEDGMENTS

We would like to acknowledge our fellow classmates from the spring 2021 Power Units and Power Trains class at Purdue University's School of Agricultural and Biological Engineering for their contributions to the structure and content of this technical chapter.

7. CONFLICTS OF INTEREST

The authors declare no conflict of interest.

8. FIELD REPAIR WELDING QUESTIONS

1. Describe how field repair welds are different from welds crafted in manufacturing facilities.
2. Describe potential problems associated with post-manufacturing repair welds.
3. Describe the conditions in which repairs may require additional preparation prior to actual welding.
4. Identify three challenges that may affect the quality of a repair weld and describe the potential solution to the issues.
5. Describe the easiest quality control check that a repair welder can make to ensure that a good weld has been applied to a broken assembly.

9. REFERENCES

[1] O. W. Blodgett, *Design of Welded Structures,* James F. Lincoln Arc Welding Foundation, 1972.

[2] American Welding Society, "Structural Welding Code—Steel (AWS D1.1)," in *Structural Welding Code,* The American Welding Society, 2020.

[3] D. H. Phillips, *Welding Engineering: An Introduction,* John Wiley & Sons, 2016.

[4] S. L. Kakani and A. Kakani, *Material Science,* New Age International Publishers, Ltd., 2012.

[5] M. June, "Mechanics of Materials-Steel," 2014. [Online]. Available: http://www.learncivilengineering.com/wp-content/themes/thesis/images/structural-engineering/Structural-steel-structural-light-gage-reinforcing.pdf. [Accessed December 18, 2021].

[6] Engineering Toolbox, "BHN—Brinell Hardness Number," 2008. [Online]. Available: https://www.engineeringtoolbox.com/bhn-brinell-hardness-number-d_1365.html. [Accessed December 18, 2021].

[7] University of Illinois, "Metals," 1995. [Online]. Available: http://matse1.matse.illinois.edu/metals/metals.html. [Accessed December 18, 2021].

[8] American Society for Testing and Materials, "About Us—ASTM," 1996. [Online]. Available: https://www.astm.org/ABOUT/overview.html. [Accessed December 18, 2021].

[9] National Board of Boiler and Pressure Vessel Inspectors, "Identifying Existing Materials," 2014. [Online]. Available: https://www.nationalboard.org/SiteDocuments/Members%20Only/Technical%20Presentations/2014-8_TechPresentation_Beach_Scribner.pdf. [Accessed December 18, 2021].

[10] MetalTekInternational, "How to Evaluate Materials," 2020. [Online]. Available: https://www.metaltek.com/blog/how-to-elevate-materials-properties-to-consider/. [Accessed December 18, 2021].

[11] O. Nguyen, "3 Most Common Industries for MIG Welding," 2018. [Online]. Available: https://www.tws.edu/blog/welding/3-most-common-industries-for-mig-welding/. [Accessed December 18, 2021].

[12] Tulsa Welding School, "What Are the Different Welding Positions?," 2020. [Online]. Available: https://www.tws.edu/blog/welding/what-are-the-different-welding-positions/. [Accessed December 18, 2021].

[13] Lincoln Global, Inc., "Parts of a Weld Poster (WC-482)," 2015. [Online]. Available: https://www.lincolnelectric.com/assets/US/EN/literature/WC482.pdf. [Accessed December 18, 2021].

[14] Miller Electric Manufacturing, LLC, "Deciphering Weld Symbols," 2007. [Online]. Available: https://www.millerwelds.com/resources/article-library/deciphering-weld-symbols. [Accessed December 18, 2021].

[15] C. D. Moran, *Interpreting Metal Fab Drawings, Salem,* Open Oregon Educational Resources, 2021.

[16] Industrial Training Partners Ltd., *Weld Defects: Causes and Corrections,* 1983.

[17] Ohio Department of Transportation, "Field Welding Inspection Guide," 2011. [Online]. Available: https://www.dot.state.oh.us/Divisions/ConstructionMgt/Materials/Miscellaneous/Field-Welding-Inspection-Guide.pdf. [Accessed December 18, 2021].

[18] American Welding Society, *Specifications for Underwater Welding (AWS D3.6M),* 2017.

CHAPTER 5. THEORETICAL STRENGTH IN BENDING

Acceptable welds are generally described as being loaded in either the transverse or the shear direction. Transverse loads are strong and described as full-strength welds. Shear loaded joints are allotted only 30% of the strength of transverse joints. These relative strength values have been used by designers for roughly 50 years, after the American Welding Society reduced originally estimated strength of shear loaded welds from 60% to 30%. No other type of joint loading is recommended for welding. This practice is almost always followed in the design and manufacture of welded products. However, there are circumstances that arise during welding repairs that may dictate that welds be loaded under other nonideal conditions. While not recommended, it is nonetheless a practical matter that some repair welds are crafted in this manner. At present, there are no guidelines for estimating the strength of these types of nonideal welds, and this work was an attempt to formulate an initial guideline. The following paper and poster were originally presented during the 2023 American Society of Agricultural and Biological Engineers Annual International Meeting, and they provide the background on developing an empirical study to determine the relative strength of single-sided welds in cantilever bending.

FILLET WELD STRENGTH ANALYSIS FOR CANTILEVER LOADING

TYLER J. MCPHERON, PURDUE UNIVERSITY AGRICULTURAL & BIOLOGICAL ENGINEERING

ROBERT M. STWALLEY III, PURDUE UNIVERSITY AGRICULTURAL & BIOLOGICAL ENGINEERING

ABSTRACT

Theoretical fillet weld strength calculations are used in the design of welded structural members to assess the strength of a welded connection in a design based on two parameters: the tensile strength of the weld filler metal and the effective area of the weld. This calculation yields a value that is independent of the mechanical properties of the base materials. The type of load applied can alter the theoretical strength of the weld considerably. This is evident in American Welding Society (AWS) Structural Welding Code D1.1, which recommends that the tensile strength of the weld filler to be reduced by 70% of its actual value for use in the calculation of a welded member configured with load applied in parallel (shear) length of the weld. This factor of 0.30 has been shown by tests in "Proposed Working Stresses for Fillet Welds in Building Construction" [1] to provide a "factor of safety ranging from 2.2 for shearing forces parallel to the longitudinal axis of the weld, to 4.6 for forces normal to the axis, under service loading" [2]. The whole value of the tensile strength of the weld filler metal is used for calculating the strength of a welded member loaded perpendicular (transverse) to the length of the weld. However, there are no similar considerations for joined materials with load applied in a bending configuration. Often in applied structural designs, fillet welded members are loaded in such a way that causes deflection of the material, thus creating a bending scenario. The cantilevered beam configuration creates a much different loading scenario with additional stresses on the weld, different from those of a perpendicular or parallel load. This research experiment aims to understand and analyze the strength of a weld using the repeatable AWS D1.1 Fillet Weld Break Destructive Test to derive a mathematical equation that yields a factor of safety in the similar range of 2.2 to 4.6 for the calculation of theoretical fillet weld strength for plates welded in a cantilever beam configuration.

KEYWORDS: bending, crush test, sample preparation, welded joint

1. INTRODUCTION

To begin the current experimental effort, typical load configurations must be introduced before further analyzing and then testing a sample piece. Simple loading conditions are divided into four respective categories: bending, axial, torsion, and transverse [3]. Shear loading a welded joint parallel to its axial line is defined as a

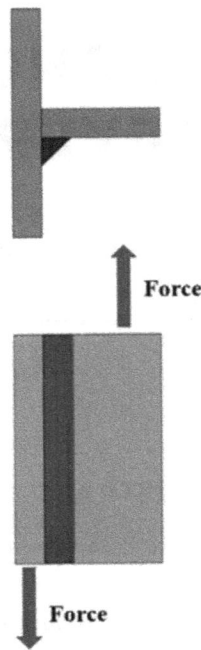

FIGURE 5.1. Parallel or shear loaded weldment.

load that is applied on the effective area of the weld in parallel with the length of the weld. A shear loaded weldment can be loaded in either tension or compression. The free body diagram for this load scenario is shown in Figure 5.1. There are additional considerations to be made when evaluating the stresses and load capacity of welds loaded in shear. This is due to the load being concentrated on the root penetration of the weld without the structural support from the whole surface area of each weld leg attached to the base metal. This is the major difference between a parallel and perpendicular loaded weldment.

Transverse load is defined as a concentrated load applied in the transversal direction, or perpendicular to the effective weld area. This is also referred to as a tensile load. When loaded perpendicular to the weld axis, the weld is often significantly stronger than welds that endure shear load. With proper root penetration achieved, a larger percentage of the weld material provides structural support to the joint in perpendicular loading. Figure 5.2 provides an example of a transversely loaded weldment.

A cantilevered beam is a structural member secured perpendicular to another structural member with one fixed end on the joint side and one free end. An example of this structure is shown in Figure 5.3. The fixed end can be constrained to the perpendicular structural member by a variety of ways such as bolted connections, welded joints, and pinned connections. This is common configuration in structural design and is used in a variety of design applications for supporting a structural load. Cantilever beams can be made up of any material. Often in structural applications, steel beam, plate, or concrete is used. The term "beam" should be thought of as an engineering term for the configuration of the structural member, not necessarily the shape and size of the material itself. A structural member is considered a beam if loaded perpendicular to its axis. Trusses are structural members that are loaded in the axial direction. Often, combinations of both trusses and beams make up the core foundation of many small engineered components to large building structures. For this research

FIGURE 5.2. Perpendicular or transverse loaded weldment.

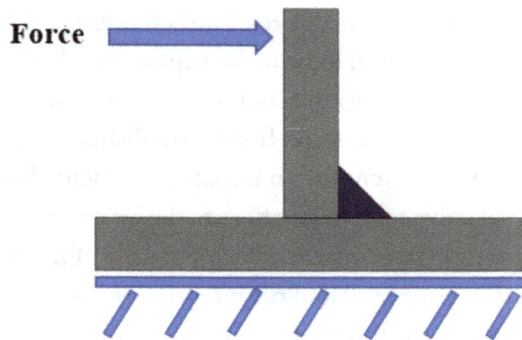

FIGURE 5.3. Cantilevered beam with one free end and one fixed end.

TABLE 5.1. *Table 2.3 from AWS D1.1 Structural Welding Code [2]*

Fillet weld	Shear on effective area	0.30 x nominal tensile strength of filler metal	Filler metal with a strength level equal to or less than matching filler metal may be used.
	Tension or compression parallel to axis of weld	Same as base metal	

experiment, a welded T-joint constructed with steel plate served as the cantilever beam. Computing the deflection of a beam with an applied load was not covered as part of the work reported here.

2. LITERATURE REVIEW

A scientific welding experiment outlined in "Proposed Working Stresses for Fillet Welds in Building Construction" [1] sought to update welding provisions and practices due to the significantly improved electrodes and steel material from those used in the structural steel design report from 1931. This revised test, conducted by a Task Committee of the AWS Structural Welding Committee, headed by Dr. A. Amirikian, chairman of the parent committee, was asked to develop a suitable program in 1969 and concluded that a more liberal working stress of 0.30 times the tensile strength of the electrode was justified, thus eradicating the previous 0.60 of the specified minimum yield stress of the steel for basic working stress used in structural steel design in 1931 [1]. These values have been used across generalized welding instruction for over 50 years [4]. The American Welding Society (AWS) D1.1 Structural Welding Code cites Higgins and Preece [1] in C2.10 of D1.1 as saying that "a working stress equal to 0.30 times the tensile strength of the filler metal, as designated by the electrode classification [5], applied to the throat of a fillet weld has been shown by tests to provide a factor of safety ranging from 2.2 for shearing forces parallel to the longitudinal axis of the weld" [2]. Table 5.1 shows the allowable stresses from AWS D1.1.

3. EXPERIMENTAL PROCEDURES

This experimental effort aimed to create a statistically relevant set of test pieces, in accordance with prior recommended testing processes, that could be used to begin an experimental exploration of fillet welds in bending. The fabrication process in manufacturing test pieces for the crush process was as controlled and uniform as possible, given the resources available for the work. It was established during the planning process that 20 samples with two weld sizes would be sufficient for an initial experiment. The AWS Fillet Weld Break Test was used to conduct a series of destructive tests in an effort to derive an allowable stress factor to be used for the weld strength design calculation to yield a consistent factor of safety for a fixed-edge cantilevered load scenario. Two plate thicknesses of 0.250 *in* (6.3 *mm*) and 0.375 *in* (9.5 *mm*) were designed to be evaluated to analyze the relationship between ultimate strength and weld size in compliance with the AWS D1.1 welding code. Materials were acquired through the Purdue University Materials Acquisition Warehouse for the following:

Test Sample A: 0.250 *in* (6.3 *mm*) A36 mild steel plate (Sample size = 20)
Test Sample B: 0.375 *in* (9.5 *mm*) A36 mild steel plate (Sample size = 20)

The statistical goal of the experiment was to record minimal standard deviation of load at failure for each plate thickness. A consistent failure load among 20 samples of each material thickness will provide the statistical basis for deriving an accurate allowable stress factor for calculating the strength of welds subject to cantilevered bending. Material thicknesses of 0.250 *in* (6.3 *mm*) and 0.375 *in* (9.5 *mm*) were specifically chosen due to the increase in weld leg size defined in AWS D1.1, as shown in Table 5.2. The percent increase in weld size

TABLE 5.2. *D1.1 Weld leg size specifications [6]*

TEST MATERIAL THICKNESS (IN)	LEG SIZE	SIZE (IN)
0.250	Minimum	0.125
0.375	Minimum	0.1875

FIGURE 5.4. A12 test specimen for fillet weld bending experiment.

will be compared with the percent increase in strength for the larger weld size of the two samples. One of the finished 0.250 *in* (6.3 *mm*) "B" test samples is shown in Figure 5.4.

The test sample manufacturing protocol and testing process for the bending experiment was defined with the following fabrication steps:

1. Weld prep and welding
    1.1 Grind off all mill scale. Weld two A36 mild steel plates together in a T-joint configuration welded on one side of the joint, as in Figure 5.5.
    1.2 Visually inspect weld for excess porosity and undercut. Void test specimen if such inclusions are visible.
    1.3 Cut a 1 *in* cross-section of the test specimen on both ends of the test specimen.
    1.4 Measure the size of the fillet weld on the now 6 *in* (15 *cm*) test specimen with a fillet weld gauge, as in Figure 5.6, or with dial calipers, shown in Figure 5.7, for a more precise measurement. The fillet weld size must be uniform for the entire weld length. Photograph and record the measurement according to test specimen number.

FIGURE 5.5. Plate configuration for welding pieces into test sections.

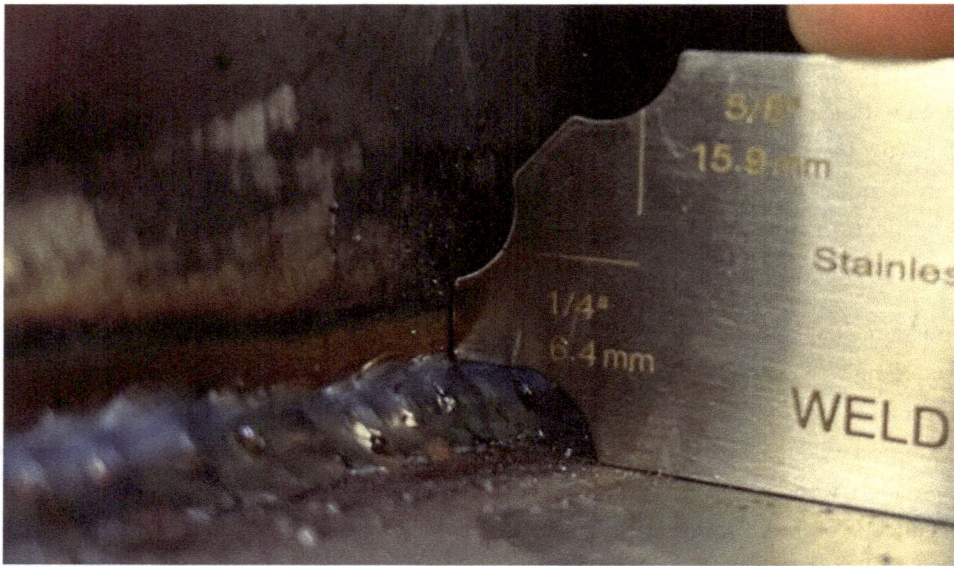

FIGURE 5.6. Measuring with fillet weld gauge.

FIGURE 5.7. Measuring with dial callipers for more precision.

2. Macroetch Test

    2.1 Grind and polish the welded joint area on one of the 1 *in* (2.5 *cm*) cross-sections.

    2.2 Apply phosphoric acid to the welded joint while the joint is still warm from polishing.

    2.3 Allow 2 min to let the phosphoric acid etch the weldment.

    2.4 Visually inspect the joint for adequate root fusion and equal fusion into the base metal. Photograph and record the root fusion.

3. Repeat steps 1.1—2.4 for 40 test specimens.

    3.1 Number each test specimen for tracking purposes. Assemble for transport to the test site, as shown in Figure 5.8.

4. Data Collection

    4.1 Install test specimen in MTS Systems (Eden Prairie, MN) Insight electromechanical press with a real-time data logger to record and plot load and displacement, oriented like Figure 5.9.

    4.2 Apply continuous load to the test specimen until the load curve is completed.

    4.3 Remove test specimen and inspect the edge of the break for root fusion in compliance with AWS D1.1 visual inspection criterion for a fillet weld break test.

5. Repeat steps 4.1–4.3 for each sample of 20 test specimens.

All of the test piece fabrication was performed at the Purdue University Agricultural and Biological Engineering Student Achievement Center. To achieve minimal deviation between samples, each test specimen was configured and welded with consistent positions and machine settings. Preliminary test T-joints were

FIGURE 5.8. Weld test specimens: Sample A on left and Sample B on right.

fabricated to adjust the welding machine for optimal performance. A macroetch test was conducted on the test T-joints to evaluate root fusion for different machine settings. The final welding machine specifications are shown in Table 5.3, with some adjustment tolerance due to building current draw and minor machine error.

Each test specimen was placed in the same position on the welding table, with the ground clamp placed in the same location for all samples. Each test specimen was welded in the 2F horizontal position with a slight support underneath the specimen for a partial 1F flat welding position. The specimen was configured this way for welder comfort. Figure 5.10 shows the test specimen setup for manufacture.

The AWS has defined criteria for the inspections necessary to complete before and after a fillet weld break test has been conducted. Before destructive testing of the test specimen, visual inspections should be completed by a certified welding inspector. The weld should be reasonably uniform throughout the length of the weld and feature no exterior cracks, excess undercut, or porosity [2]. Assuming that the test specimen clears the visual inspections, the test is considered a pass even if the specimen bends flat upon itself. Many think that it is a failed test if the weld fractures. However, if the weld fractures on the centerline of the weld throat, shows complete fusion to the joining base metals, and shows no inclusion or porosity larger than 3/32 *in* (2.5 *mm*) in its greatest dimension and the sum of the greatest dimensions of all inclusions and porosity do not exceed

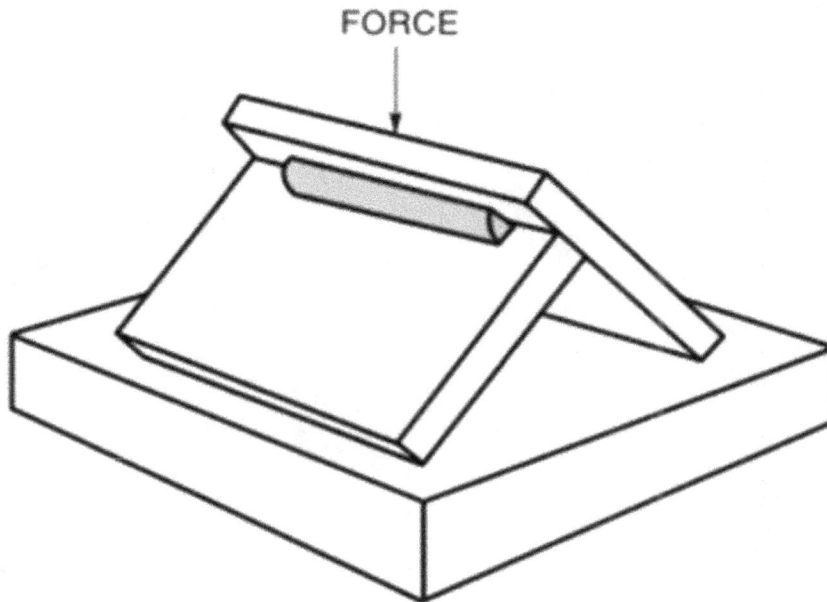

FIGURE 5.9. Fillet weld break test configuration [6].

TABLE 5.3. *Welding equipment specifications*

WELDER	MILLER MILLERMATIC 252 MIG WELDER	230V
Voltage (V)	19.8 ± 0.5 (for ¼ *in* material)	21.9 ± 0.5 (for 0.375 *in* material)
Wire Feed Speed (*in/min***)**	405 ± 10 (for ¼ *in* material)	410 ± 10 (for 0.375 *in* material)
Shielding Gas	25% Argon, 75% CO_2	—
Filler Metal	E70S Filler Wire (70,000 *psi*)	—
Filler Metal Diameter	0.035 *in*	

0.375 *in* (9.5 *mm*) in the 6 *in* (15 *cm*) long specimen, the test specimen is considered to have passed [2]. It should be noted that it is common for the fillet weld to fracture during this test, especially if the base material is ½ *in* (12.7 *mm*) or thicker [7].

A macroetch test was a test conducted by applying an acidic metal etching chemical to a cross-section of a fillet weld. A variety of chemicals can be used to achieve the etched results. In this experiment, phosphoric acid will be used to macroetch test each test specimen. Many welds can look uniform and structurally sound on the exterior but may have hidden imperfections underneath. The macroetch test is used to ensure that proper root fusion into the joining material is achieved. The AWS specifies that the weld shall show fusion to the root of the joint but not necessarily in excess amounts [2]. Excessive root fusion can weaken the joint significantly and cause unacceptable amounts of undercut. This usually indicates too much current on the

FIGURE 5.10. Welded test specimen ready for initial inspection.

machine settings or a slow electrode travel speed. The proper leg size should also be measured at the time of the macroetch test.

There are certain assumptions necessary to complete the logic chain for the bend testing analysis. The purpose for using the AWS fillet weld break test is to validate the data collected by implementing standard welding procedures while following a standardized testing procedure to validate the 20 samples of 0.250 *in* (6.3 *mm*) T-joint test specimens and 20 samples of 0.375 *in* (9.3 *mm*) T-joint test specimens. This procedure will validate the data collected and provides the basis for a scientifically binding experiment. However, the assumption that a noncertified welder fabricated the test specimens for the experiment must be known. Additional considerations are to follow proper procedures, measure necessary parameters outlined by AWS testing criteria, and be willing to exclude any noncompliant test specimens from the results to ensure the validity of the data collected in the experiment. In comparison, the Task Committee of the AWS Structural Welding Committee in 1968 sent test specimens to a variety of different certified welders in different regions in the United States. The ideal scenario would be to program a robotic welder to weld all the test specimens used in the experiment, but resources for this experimentation were limited.

4. DATA COLLECTION AND TESTING

Table 5.4 details the equipment used in this experiment. An MTS Systems (Eden Prairie, MN) Insight electromechanical press, shown in Figure 5.11, was the primary piece of equipment used to carry out the experiment. The MTS press had a woo *kd* wapacdty (67,400 *lbf*). A preliminary finite element analysis study was conducted to estimate the amount of force required to break each test specimen or achieve the ultimate strength on the stress/strain curve. The finite element analysis results showed that the joint would fail at

TABLE 5.4. *Equipment for data collection and testing welded samples in bending*

Description	Mfg./Model	Specification
Electromechanical Press	MTS Insight	300 kN Standard Length
Load Cell	MTS—569331-01	300 kN Capacity
Software	Labworks	—

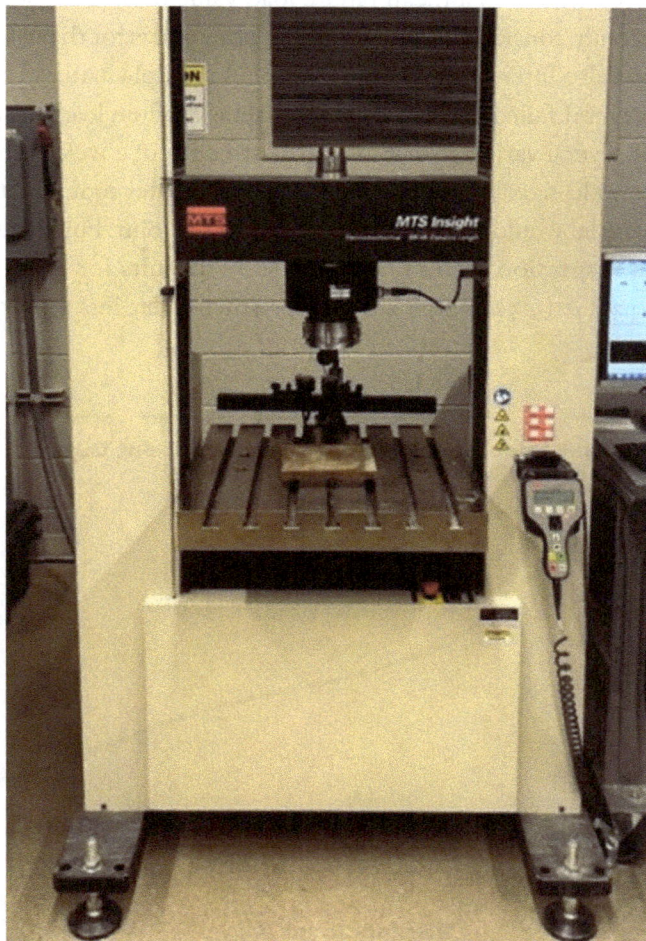

FIGURE 5.11. MTS Systems (Eden Prairie, MN) Insight electromechanical press.

approximately 5,000 lb_f (22.2 kN). This ensured that the MTS press would be sufficient for testing the fillet weld specimens. The Labworks (Lehi, UT) software paired with the MTS press was capable of actively plotting the stress/strain curve for each sample. This provided an excellent amount of data to collect measured values for each test specimen to meet the test objectives.

Measured values for each test specimen include load, time, extension, stress, and strain. Each test specimen was fixtured in the MTS machine in the same orientation, matching the AWS fillet weld break test loading criteria. After fixturing the test specimen in the machine, the press head was lowered to slightly contact the test specimen, then zeroed using the software program before beginning the test. Beginning the test initializes an applied continuous load to the test specimen. These tests produced the ultimate strength curve for each test specimen, or the maximum amount of load each test specimen can absorb by stretching or pulling before fracture [3]. Figure 5.12 presents the results for test specimen A7.

The ultimate strength was typically reached within the first few seconds of initializing the test. The test was stopped once the load declined to 1,000 lb_f (4.4 kN). These steps were repeated for 20–0.250 in (6.3 mm) thick and 20–0.375 in (9.3 mm) thick test specimens. Figure 5.13 shows a crush test in progress.

The failure load results were recorded for each test specimen. Consistent results were achieved for the 0.250 in (6.3 mm) thick A test specimens with a mean failure load value of 6,640 lb_f and a standard deviation of 909 lb_f. This was a desirable result, considering the amount of potential error that can occur by manually welding each test specimen. It was also interesting to note that the A7 sample featured a slightly smaller weld size than the rest of the samples, yet it failed above the sample average failure load at 7,010 lb_f. This was an initial indication that there can be several variables that affect the strength of a weld rather than just the effects of increasing or decreasing the weld size. It could be that there is a greater root penetration, producing a small leg size, since more weld material would be concentrated within the joint. Further analysis will be conducted in the fillet weld break test inspection section. Table 5.5 shows results for 0.250 in (6.3 mm) thick A test specimens. Results for the 0.375 in (9.3 mm) thick B test specimens were less uniform than the results for the

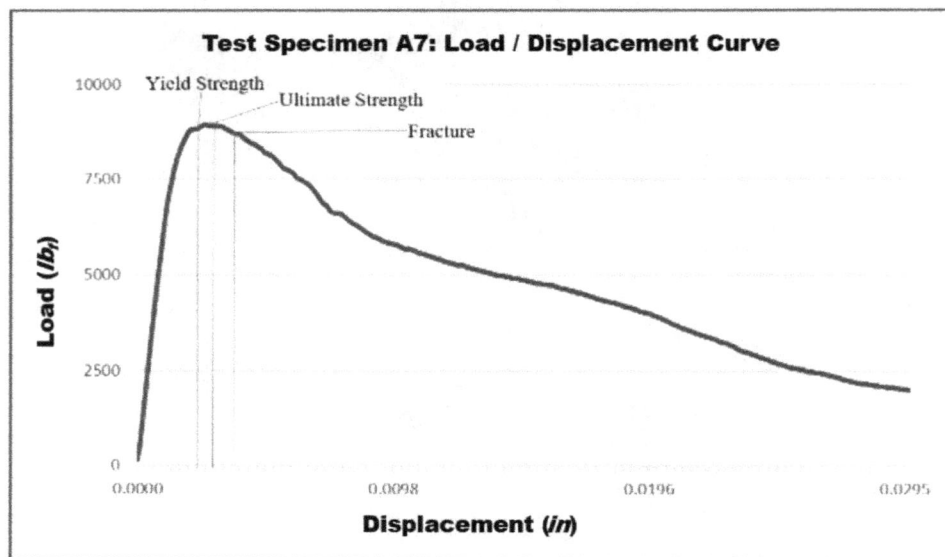

FIGURE 5.12. Stress/strain curve for A7 test specimen in bending experiment.

FIGURE 5.13. Test specimen in MTS Systems (Eden Prairie, MN) Insight electromechanical press.

0.250 *in* (6.3 *mm*) thick test specimens. However, for the size of the sample, the results will suffice. Again, the potential error that exists due to manually welding and preparing each test sample creates a large variability in results. The mean failure load for sample B was 12,400 lb_f, with a standard deviation of 1,600 lb_f. Even though the average size of the weld for B sample was only 12% higher than the average weld size for A samples, a 60% increase in ultimate load capacity was observed for the larger pieces. Table 5.5 presents these results. It is important to clarify that the data above represent the raw, unfiltered data for the experiment. This data has not yet been cleaned and evaluated for sample failures and outliers that could potentially skew results. Therefore, the above data in their entirety will not be used for objective statistical experiments at this time.

The MTS electromechanical press did not break the test specimens apart during testing. Because the load was lifted from each test specimen once it decreased to 1,000 lb_f (4.4 *kN*), if the test specimen were to break before the end of the load curve this would indicate a very weak weld with little penetration. Each T-joint was later separated manually for further inspection. The main objective for the fillet weld break test is to inspect root penetration at the joint and distribution of weld metal to the base metals. This is completed by separating the T-joint and inspecting the break. A clean break directly down the center axis of the weld indicates that there is equal distribution of weld metal on both joining materials. This also indicates that root penetration is symmetric and not favoring one side of the joint or the other. Favoring either side is an indication that

TABLE 5.5. *Measured data summary statistics*

SAMPLE	MATERIAL THICKNESS (IN)	μ LEG SIZE (IN)	μ THROAT SIZE (IN)	μ FAILURE LOAD (LB)	FAILURE LOAD STD. DEV. (LB)
A	0.25	0.247	0.175	6,641	909
B	0.375	0.278	0.196	12,371	1,603

TABLE 5.6. *AWS fillet weld break test acceptance criteria* [2]

THE BROKEN SPECIMEN SHALL PASS IF	
(1)	The specimen bends flat upon itself.
(2)	The fillet weld, if fractured, has a fracture surface showing complete fusion to the root of the joint.
(3)	No inclusions or porosity larger than 3/32 in (2.5 mm) in its greatest dimension.
(4)	The sum of the greatest dimensions of all inclusions and porosity shall not exceed 0.375 *in* (10 *mm*) in the 6 *in* (15 *cm*) long specimen.

electrode angle is either too steep or too shallow in relation to the joint at that point. Table 5.6 presents the overall weld failure points for specimen testing.

5. STATISTICAL ANALYSIS

To analyze the results of the experiment, four main statistical tests have been performed. Each research question was articulated to analyze a few key objectives of the experiment. These tests, along with their hypothesis, are shown in Table 5.7. There is enough data to do multiple analysis relating to the stress and strain curves produced when loading each sample. However, the focus for this experiment was on the ultimate strength of the weld in failure loading, the effective area of the weld, and the relationship between calculated theoretical strength and measured ultimate strength. All statistical tests were performed using R programming language for statistical computing.

6. SUMMARY OF FINDINGS

Considering all the test specimens from both samples, no significant outliers were discovered due to imperfections or impurities within the weld. None of the 13 test specimens that failed the post-break test inspection were outliers in the data, when considering the point at which failure occurred. Since the measurements were consistent between the passing test specimens and the failed test specimens, the same conclusions would have likely been drawn from the statistical tests using either the cleaned data or the raw data. Alternatively, welds that lacked large amounts of root fusion or had large slag inclusions and porosity would have likely showed underperforming results compared to results achieved by sufficient welds. Many of the welds that failed inspection had a lack of root fusion into the base metal. Of the 40 test specimens, one failed visual inspection before destructive testing. Since 27 out of 39 test specimens passed inspection after destructive testing, this

TABLE 5.7. *Statistical analysis summary*

STATISTICAL OVERVIEW

1. Failure Load vs. Effective Weld Area

Is there a difference between failure load (*lb_f*) of 0.25″ tee joints (sample A) and 0.375″ tee joints (sample B) due to the difference in required weld sizes for the two base material thicknesses?

    H_o: Sample A failure load = Sample B failure load

    H_a: Sample A failure load < Sample B failure load

    Test: Two-Sample t-Test

Are there any significant outliers for failure load (*lb_f*) due to possible impurities in the weld for each sample of T-joints?

    H_o: Significant outliers for failure load for both samples = 0

    Test: Identify outliers

2. Allowable Stress and Design Factor

Is there a difference between calculated equation factors between sample A and sample B?

    H_o: μ equation factor for Sample A = μ equation factor for Sample B

    H_a: μ equation factor for Sample A ≠ μ equation factor for Sample B

    Test: Two-sample t-Test

Can an accurate conclusion be found as an acceptable equation factor for use in the design of components featuring welded joints loaded in a cantilevered bending configuration?

    H_o: μ difference of equation factor between Sample A and Sample B = 0

    Test: Comparison of means

proves that visually appealing welds do not guarantee structural integrity. This experiment showed that welds with uniform and aesthetic appearance may not possess all the critical requirements for a weld that is structurally sound.

Failure loads measured for each sample yielded drastically different values, with consistent results measured from each sample. The T-joint test specimens constructed with thicker 0.375 *in* (9.3 *mm*) A36 steel material showed an average failure load twice the average failure load of the test specimens constructed with 0.250 *in* (6.3 *mm*) material. The effective weld area was an average difference of 0.14 *in²* (0.92 *cm²*) between samples. The difficulty in telling how much difference in effective weld area between the two samples and the difference in base metal thickness influenced the ultimate load capacity was discussed. Due to the loading configuration, there could have been additional structural stability derived from the thicker materials, allowing the welded connection to withstand greater loads. The larger weld size, if properly constructed, does increase the amount of fusion into the base metals and will certainly increase the strength of the connection [8]. However, it is difficult to quantify the influence of these parameters between two samples with two different plate thicknesses requiring different weld sizes to comply with AWS D1.1 weld size specifications.

Using similar theory and processes used in "Proposed Working Stresses for Fillet Welds in Building Construction" [1], two allowable stress factors were derived for both test samples. An average factor of 0.10 was calculated for the 0.250 *in* (6.3 *mm*) test specimens in Sample A, and an average factor of 0.15 was calculated for the 0.375 *in* (9.3 *mm*) test specimens in Sample B. This factor derived from the 1968 experiment is multiplied by the ultimate tensile strength of the weld filler metal used in the welding process to reduce its value by 70%. In this experiment, ultimate tensile strength reductions of 90% for Sample A and 85% for Sample B were calculated by using the ultimate load capacity measured in the experiment and the effective weld area of each test specimen. It can be concluded that welds loaded in cantilevered bending configuration experience more stresses and are subject to failure at loading premature to that of shear load configurations and certainly transversely loaded fillet welds, which use the total rating of weld filler metal tensile strength for evaluating load capacity of a weld.

7. THEORETICAL WELD BEND STRENGTH QUESTIONS

1. Describe the original experimentation that determined the relative strength of shear loaded welds versus transverse loaded welds.
2. What statistical controls were in place for the AWS testing that developed the modern reduction in shear weld strength guideline?
3. What limitations in the design of the present experimental program curtail its general applicability?
4. Describe the weld pass/fail criteria in the bending experiment testing.
5. Why was it necessary to examine two base material section thicknesses in the experiment?

8. REFERENCES

[2] American Welding Society, *AWS D1.1 Structural Welding Code—Steel*. American Welding Society, 2000.

[6] American Welding Society, *AWS D1.1 Structural Welding Code—Steel*. American Welding Society, 2020.

[4] Blodgett, O. W., *Design of Welded Structures*. The James F. Lincoln Arc Welding Foundation, 1972.

[5] Canadian Welding Bureau, "Electrode Certifications," 2021. Retrieved from CWB Group: https://www.cwbgroup.org/safety/industry/electrode-certification.

[3] Krutz, G. W., J. K. Schuller, and P. W. Claar, *Machine Design for Mobile and Industrial Applications*. Society of Automotive Engineers, Inc., 1994.

[8] SEU, "Welding Myths Debunked—Maximum Fillet Weld Thickness," 2017. Retrieved from Structural Engineering University: https://learnwithseu.com/welding-myths-debunked-maximum-fillet-weld-thickness/.

[1] Higgins, T. R., and F. R. Preece, "Proposed Working Stresses for Fillet Welds in Building Construction," *AISC Engineering Journal*, vol. 6, pp. 16–20, 1969.

[7] Zuheir Barsoum, M. K., "Ultimate Strength Capacity of Welded Joints in High Strength Steels," *Procedia Structural Integrity*, vol. 5, 1401–1408, 2017.

PURDUE UNIVERSITY®

Agricultural and Biological Engineering

Fillet Weld Strength Analysis for Cantilever Loading

Tyler J. McPheron & Dr. Robert M. Stwalley III

Introduction

This research experiment aims to initially understand and analyze the strength of a weld in cantilevered bending using the repeatable AWS D1.1 Fillet Weld Break Destructive Test to derive a mathematical equation that yields an allowable stress factor for the calculation of theoretical fillet weld strength for plates welded in a cantilever beam configuration.

Loading Conditions
- Shear Load
- Transverse Load
- Bending Load

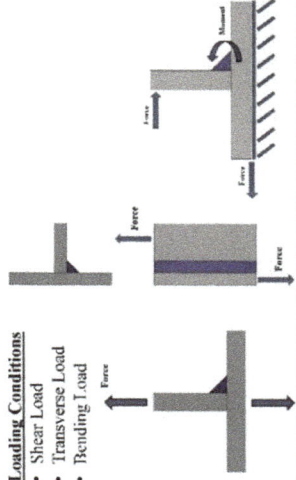

Experimental Procedure

AWS D1.1 Fillet Weld Break Destructive Test

Sample A: 0.250-in A36 Mild Steel (n=20)
Sample B: 0.375-in A36 Mild Steel (n=20)

- One-sided fillet welded tee-joint
- Visual inspection
- Macroetch root fusion inspection
- Weld leg size measurements

Data Collection & Testing

Equipment & Test Configuration

- MTS Insight Electromechanical Press 300 kN Capacity
- Test specimens loaded weld side up, following AWS procedures
- Continuous load applied until load curve decreased to 1,000 lb_f or less
- Weld quality and root fusion inspected after bending

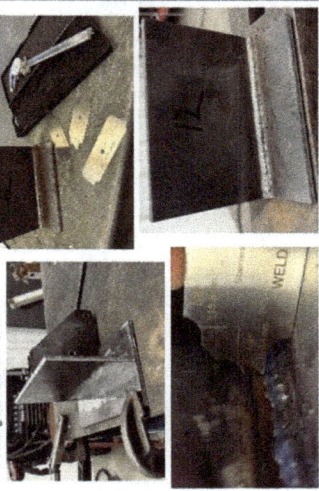

Test Specimen A7: Stress/Strain Curve

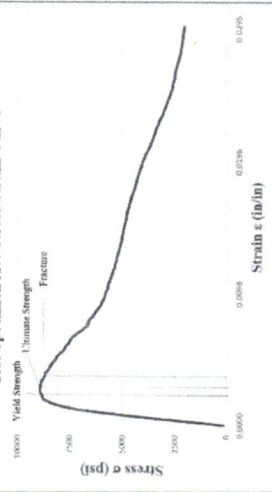

Data Collection
- Load, time, extension, stress, and strain
- Active plot of stress/strain curve for each test specimen
- All measured values for each test specimen exported to csv files
- Ultimate strength measured for each test specimen

Statistical Analysis

Failure Load vs. Effective Weld Area

H_0: Sample A mean failure load = Sample B mean failure load
H_a: Sample A mean failure load < Sample B mean failure load

Conclusion: Reject null hypothesis

Two-Sample T-Test

Sample	Mean	Std Dev	P-Value	t-Value	DF	Min	Max
A	6,870 lb_f	774 lb_f	p<0.001	-11.81	25	5,930 lb_f	8,411 lb_f
B	12,546 lb_f	1,654 lb_f	p<0.001	-11.81	25	9,474 lb_f	15,360 lb_f

Allowable Stress & Design Factor

H_0: μ equation factor for sample A = μ equation factor for sample B
H_a: μ equation factor for sample A / μ equation factor for sample B

Conclusion: Rejected null hypothesis

Two-Sample T-Test

Sample	Mean	Std Dev	P-Value	t-Value	DF	Min	Max
A	0.10	0.01	p<0.001	-7.97	25	0.08	0.13
B	0.15	0.02	p<0.001	-7.97	25	0.11	0.19

$$E_f = \frac{P_{ultimate}}{\sigma_t * A}$$

E_f = allowable stress reduction factor
$P_{ultimate}$ = failure load from experiment (ultimate strength)
A = effective weld area (in²)
σ_t = ultimate tensile strength of weld filler (psi)

Summary of Findings

Ultimate Strength
- The slightly larger effective weld area on the 0.375-in tee-joints yielded greater ultimate strength than the 0.250-in tee-joints

Allowable Stress Factor
- 0.10 and 0.15 stress factors calculated for both samples are much lower than the AWS 0.30 factor used for shear loaded welds

Structural Welding Observations
- Double sided fillet welds yield superior strength
- Visually aesthetic welds do not indicate a properly welded connection with sufficient root penetration

References:

Acknowledgements:
Scott Brand – Agricultural & Biological Engineering, Purdue University
Jeffrey Lynch – Research Machining Services, Purdue University

CHAPTER 6. TESTING OF WELDS IN BENDING

This chapter provides more detail regarding the actual experiment into cantilever loaded single-sided welds. The details of the experiment and the statistical process used to analyze the data are provided. The data for failure versus effective weld area and allowable stress versus acceptable design factors were reviewed. The overall goal of this effort was to develop a preliminary guideline for the appropriate level of strength reduction to apply to a single-sided weld when it was subjected to cantilevered loading. Although this was only an initial study, which would require far more extensive planning and samples to definitively provide a design strength reduction factor, it was clearly shown that welds of this type are weak. The effective strength reduction found in this experiment was between 85% and 90%, confirming the general recommendation not to create this general style of welded joint if at all possible. The essay below was originally published in the *Journal of Welding Technology*.

FILLET WELD STRENGTH ANALYSIS FOR CANTILEVER LOADING: AN INVESTIGATION OF SINGLE-SIDED FILLET WELD STRENGTH IN BENDING APPLICATIONS

TYLER J. MCPHERON, DEPARTMENT OF AGRICULTURAL & BIOLOGICAL ENGINEERING, PURDUE UNIVERSITY

ROBERT M. STWALLEY III, DEPARTMENT OF AGRICULTURAL & BIOLOGICAL ENGINEERING, PURDUE UNIVERSITY

ABSTRACT

Theoretical fillet weld strength calculations to assess the strength of a welded connection in a design are based on two parameters: the tensile strength of the weld filler metal and the effective area. The type of load applied can alter the theoretical strength of the weld considerably. American Welding Society Structural Welding Code D1.1 recommends a reduction by 70% for use in a welded member configured with load applied in parallel with the weld. This remaining factor of 0.30 has been shown by well-accepted tests to provide factors of safety between 2.2 for shearing forces parallel to the longitudinal axis of the weld and 4.6 for forces normal to the axis, under service loading. The whole value of the tensile strength of the weld filler metal is used for calculating the strength of a welded member loaded perpendicular to the weld. However, there are no similar considerations for a load applied in a bending configuration. While not a recommended design in applied structural designs, fillet welded members can be loaded in a way that causes deflection of the material, thus creating a bending scenario. The cantilevered beam configuration creates a much different loading scenario with additional stresses on the weld, different from those of a perpendicular or parallel load, and can occur particularly in repairs. This research experiment was to initially understand and analyze the strength of a gas metal arc weld in cantilevered bending and to derive a mathematical equation that yields a factor of safety in the similar range of 2.2–4.6.

KEYWORDS: bending, cantilever loading, crush test, sample preparation, welded joint Nomenclature:

A = effective weld area (in^2)

E_f = allowable stress factor/equation factor

F_t = weld strength (lb_f)

L_s = leg size (*in*)

L_w = weld length (*in*)

P_f = failure load from experiment (ultimate strength [lb_f])

W_t = theoretical throat thickness (*in*)

σ_t = ultimate tensile strength of weld filler (*psi*)

1. INTRODUCTION

The care and planning taken in the engineering design of a weldment for a mechanism may not always be present in the field repair of the same device, or the individual performing the repairs may be forced into a situation where the lesser of two evils necessitates doing something not optimal. The needed welding repair may not be in a position where both sides are accessible, or the repair personnel may not recognize the importance of a structural joint with good penetration on both sides of the repair. In either event, where only one edge of the two pieces manages to get attached, single-sided welds can be produced occasionally, and they can ultimately carry cantilevered loads. Even though the scenario is less than desirable and is not suggested, it would be appropriate if a generalized strength reduction for the situation were known. However, little seems to be published for this state of affairs other than to say that it is not recommended. Therefore, the research objective in this effort will be to determine an initial quantifiable reduction in strength for a single-sided weld in a cantilever loaded situation. The balance of this essay will consist of a section covering the background on the experiment and a literature review focusing on the subject, the methodology of the experimentation and the materials used to execute the work, the results of the testing and a discussion of the outcomes, and the conclusions about this effort and recommendations for further study.

2. BACKGROUND AND REVIEW

Many established machine design texts include sections on welding processes [1–4]. They tend to concentrate on basic welded joints and ignore nonstandard circumstances such as those under consideration here. An understanding of various load configurations is provided in this section as background information to further aid in the analysis and testing of a bending load. This section of the essay will also examine prior investigative work done on cantilever loads and offer known information about limiting considerations of the study.

2.1 BACKGROUND

Simple loading conditions are divided into four respective categories: bending, axial, torsion, and transverse [4]. A shear loaded welded joint is defined by a force that is applied across the effective area of the weld in parallel with the length of the weld. A shear loaded weldment can be placed in either tension or compression. The free body diagram for this load scenario is shown in Figure 6.1. There are additional considerations when evaluating the stresses and load capacity of welds loaded in shear. This is due to the load being concentrated on the root penetration of the weld, without the structural support from the whole surface area of each weld

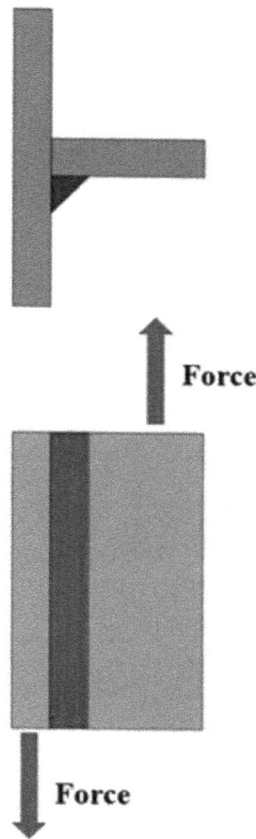

FIGURE 6.1. Loads present in parallel or shear loaded weldments.

leg attached to the base metal. This is the major difference between a parallel and a perpendicular loaded weldment.

Transverse load is defined as a force applied in the transversal direction, or perpendicular to the effective weld area. This is also referred to as a tensile load. When loaded perpendicular to the weld axis, the weld is often significantly stronger than welds that endure shear load. When proper root penetration is achieved, a larger percentage of the weld material provides structural support to the joint in perpendicular loading. Figure 6.2 provides an example of a transversely loaded weldment.

A cantilevered beam is a structural member secured perpendicular to another structural member with one fixed end on the joint side and one free end, similar to the free body diagram in Figure 6.3. A fixed end can be a perpendicularly constrained structural member by a variety of means, such as bolted connections, pinned connections, or welded joints. This is a common configuration in structural design and is used in a variety of design applications for supporting a structural load. Cantilever beams can be made of any material. Often in structural applications, steel beams, plates, or concrete are used. The term "beam" should be thought of as an engineering term for the configuration of the structural member, not necessarily the shape and size of the material itself. A structural member is considered a beam if loaded perpendicular to its axis.

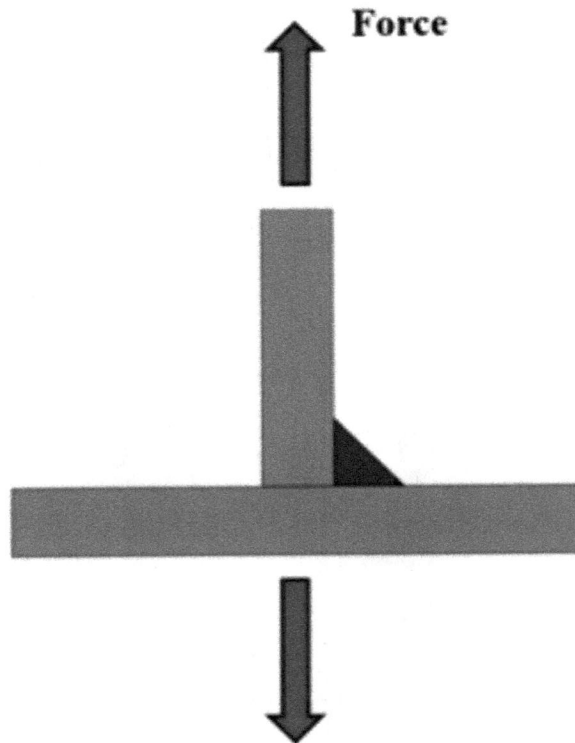

FIGURE 6.2. Loads present in perpendicular or transverse loaded weldment.

Trusses are structural members that are loaded in the axial direction. Often, combinations of both trusses and beams make up the core foundation of many small engineered components as well as large building structures. While not recommended for welded joint design, this cantilevered configuration can be present in many field repairs, where it is impossible to get to the opposite side of the weldment [5]. This effort was intended to provide quantitative information on how structural components with single-sided welded joints would handle cantilevered loading. For this research experiment, a welded T-joint was constructed with steel plate serving as the cantilever beam. McPheron and Stwalley [6] previously reported on the design of this experiment. Computing deflections for the test pieces with an applied load was not part of the scope of the current work.

2.2 LITERATURE REVIEW

As expected, since this is a non-ideal joint, preliminary examination of the literature does not find many references to the specific cantilevered configuration under investigation in this study. Gomez et al. [7] provide a good primer on out-of-plane bending and, as shown in Figure 6.4, categorize common loading scenarios for bending as "in-plane" and "out-of-plane." There has been some experimentation with the out-of-plane scenarios, and design recommendations for this configuration exist [8]. However, as shown in Figure 6.5, the

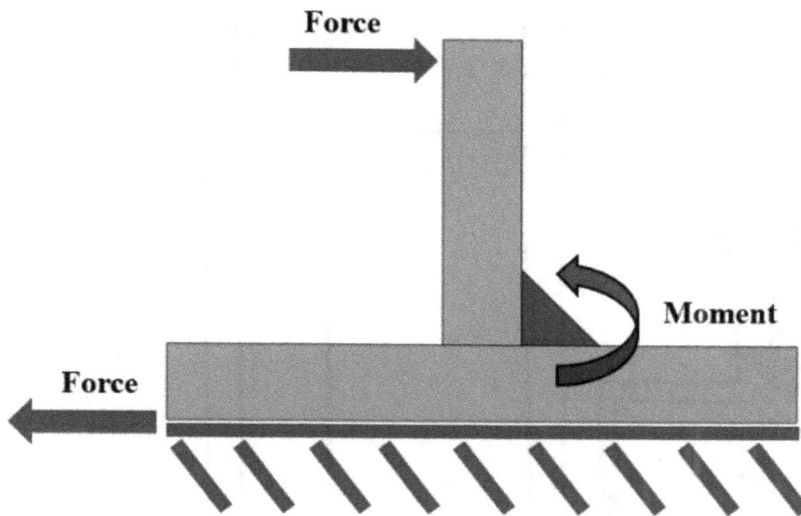

FIGURE 6.3. Loads present in a cantilevered beam with one free end and one fixed end.

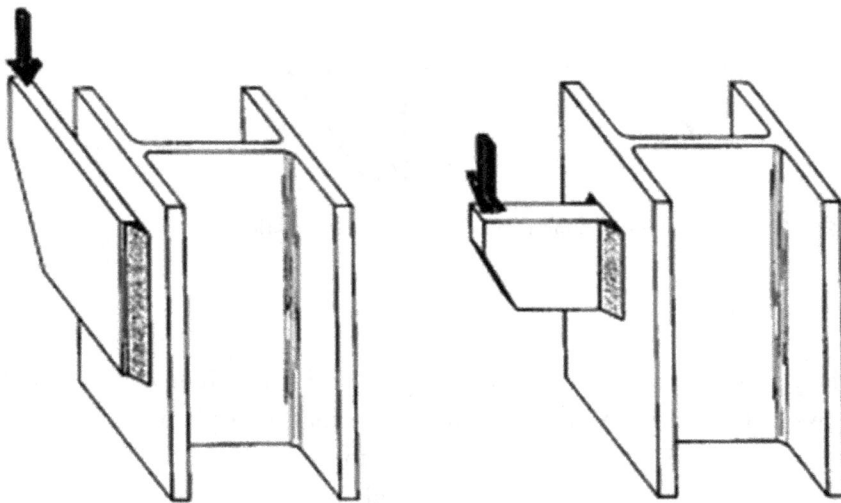

FIGURE 6.4. Type of rotation induced by eccentric loads, in-plane on left and out-of-plane on right.

current bending scenario under investigation could also be described as a combined load and torque placed onto the weld joint, and the weldment will have difficulties resisting deformation and failure under load. Clearly, the lack of exposition in the literature on this type of weld is due to the disadvantaged nature of the joint and its resulting valid criticisms [9-11].

Where the cantilevered scenario is referenced in the literature, it is always with members having a deeper section than a simple solid bar or flange. Some work on square tubing and round tubing attached to channel pieces and placed in bending has been published [12–16] as well as in studies on deck panels welded to bridge ribs, which are subjected to cantilevered bending [17–18]. Work has also been performed

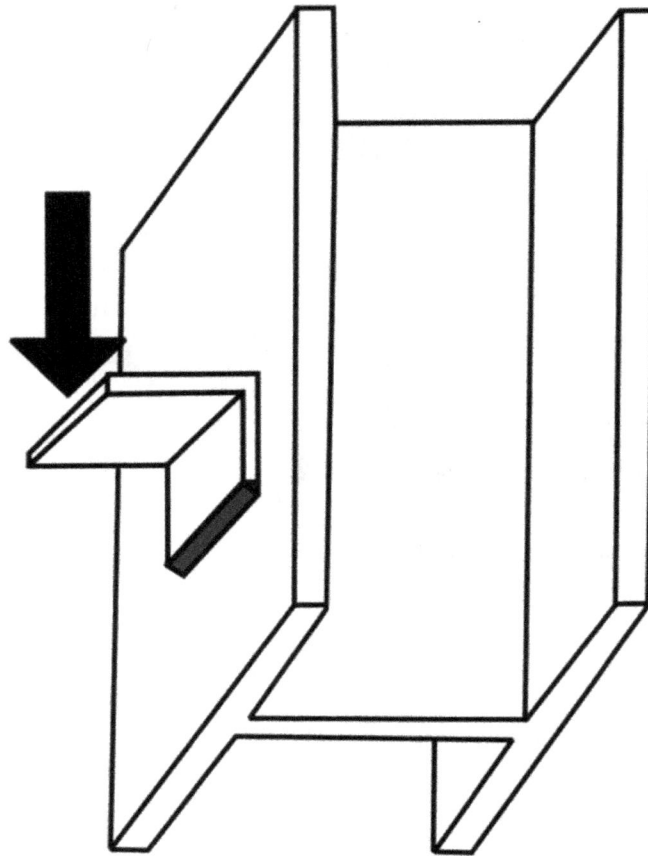

FIGURE 6.5. Out-of-plane rotation induced by a cantilevered scenario.

on strengthening this type of connection through the addition of small circle and diamond stiffeners [19]. From the section views in Figures 6.3 and 6.5, it can be seen that multiple problematic elements within standard welded construction will adversely affect any cantilevered welded joint. Welding has been demonstrated to affect the microstructure of the base materials in a joint [20], and imperfections in the welding process can certainly result in voids or inclusions, creating a stress concentration within the weld particularly with an inexperienced or poor welder. Stress concentrations in mechanical designs can be present for any number of unusual design or manufacturing features, including weldments [21–25]. Any type of stress concentration near the cantilevered joint could therefore potentially have an amplified effect, given the disadvantaged nature of this joint. That would adversely affect the long-term life of the weldment and the overall structure [26].

While not directly influencing this study, it should be noted that welded joints are also susceptible to fatigue failure [27], and cantilevered joints by their nature are particularly vulnerable. The quality of a weld has been shown to dramatically affect the performance of that weld in repeated bending load conditions [28–29]. Multiple factors have been shown to affect weld quality [30–31]. Welding current and the electrode angle have been shown to have a measurable effect on the uniformity of root depth [32–34], and inadequate penetration reduces the strength of a weld dramatically [35–36]. All factors reported in the literature that adversely affect weld strength in general are likely to have amplified effects in the cantilever loading scenario. A shallow root depth for the weld provides less weld area and leverage to resist bending. A poor weld creates imperfections which can enhance stress concentrations within or near the weld, augmenting the stresses inside the disadvantaged zone carrying the loads. This study was intended as an investigation to understand the level of that disability and to provide some quantitative guidance regarding the level of strength reduction that can be expected under this cantilevered bending scenario.

2.3 OBJECTIVES

This work was prompted by the lack of guidelines for field repair conditions that sometimes necessitate creating weldments that would not be utilized in new designs or in existing manufactured products. The experimentation was designed using historic investigations regarding weldment strength as reference [37–40]. Two sets of test pieces fabricated from two thicknesses of mild plate steel were planned as experimental samples for weld bending testing. The objectives of this study were to

1. Determine if uniform, high-quality test pieces could be fabricated for a single source sufficient for the welding bend test;
2. Determine if the thicker test pieces could bear a larger load than the thinner pieces;
3. Determine the load reduction factor for the test pieces in a cantilevered bending configuration; and
4. Determine if the load reduction factor was uniform between the two thicknesses of test pieces.

3. METHODOLOGY AND MATERIALS

Due to the significantly improved electrodes and steel material that became available during the late 1960s, scientific welding experimentation was conducted to update welding provisions and practices [37] from those used in the original structural steel design report from the 1930s [38]. The revised test, headed by Dr. A. Amirikian and conducted by a Task Committee of the American Welding Society (AWS) Structural Welding Committee, developed a suitable set of guidelines in 1968 [37]. It was concluded that a more conservative working stress of 0.30 times the tensile strength of the electrode material was justified for shear loading. This new recommendation revised the previous 0.60 value of the specified minimum yield stress of the steel for basic working stress used in structural steel design in 1931. The AWS D1.1 Structural Welding Code cites Higgins and Preece [37] in C2.10 of D1.1 says that "a working stress equal to 0.30 times the tensile strength of the filler metal, as designated by the electrode classification, applied to the throat of a fillet weld has been shown by tests to provide a factor of safety ranging from 2.2 for shearing forces parallel to the longitudinal axis of the weld." Table 6.1 lists the allowable stresses from AWS D1.1 Table 2.3 [39, 40]. Since no such guidelines exist for the cantilever bending scenario, this experimentation seeks to provide some preliminary guidance. The balance of this section will describe the test piece fabrication process for the experimentation, the bending test procedure, the inspections, and the assumptions within the current investigation.

TABLE 6.1. *Table 2.3 from AWS D1.1 [39, 40]*

Fillet weld	Shear on Effective Area	0.30 x Nominal Tensile Strength of Filler Metal	Filler Metal with a Strength Level Equal to or Less Than Matching Filler Metal May Be Used.
	Tension or compression parallel to axis of weld	Same as base metal	

3.1 EXPERIMENTAL PROCEDURES

The current experimental effort aimed to create a statistically relevant set of gas metal arc weld test pieces, in accordance with recommended testing processes, that could be used to begin an experimental exploration of fillet welds in bending. The fabrication process manufacturing the sample pieces for the crush testing was as controlled and uniform as possible, given the resources available for the work. It was established during the planning process that 20 samples with two weld sizes would be sufficient for an initial experiment. The AWS Fillet Weld Break Test [41] was used to conduct a series of destructive tests in an effort to derive the allowable stress factor to be used for the weld strength design calculations to yield a consistent factor of safety for a fixed edge, cantilevered load scenario. Two plate thicknesses of 0.250 *in* (6.3 *mm*) and 0.375 *in* (9.5 *mm*) were evaluated by analyzing the relationship between ultimate strength and weld size in compliance with the AWS D1.1 welding code. Materials were acquired through the Purdue University Materials Acquisition Warehouse for the following text samples:

Test Sample A: 0.250 *in* (6.3 *mm*) A36 Mild Steel Plate (sample size = 20)
Test Sample B: 0.375 *in* (9.5 *mm*) A36 Mild Steel Plate (sample size = 20)

The statistical goal of the experiment was to record data from enough samples to calculate the standard deviation of the load at failure for each plate thickness. A consistent failure load among the 20 samples of each material thickness would provide the statistical basis for deriving an accurate allowable stress reduction factor for calculating the strength of welds subject to cantilevered bending. Material thicknesses of 0.250 *in* (6.3 *mm*) and 0.375 *in* (9.5 *mm*) were specifically chosen due to the increase in weld leg size defined in AWS D1.1, as shown in Table 6.2 for the larger thickness plate. The percent increase in weld size will be compared with the percent increase in strength for the larger weld size of the two samples. One of the finished 0.250 *in* (6.3 *mm*) "A" test samples is shown in Figure 6.6.

The test sample manufacturing protocol and testing process for the bending experiment was defined with the following fabrication steps:

1. Weld Preparation and Welding
   1.1 Grind off all mill scale. Mark all component pieces for assembly, and clamp them to the welding surface in the same manner using C-clamps and lockjaw pliers. Electrically ground the pieces in the same position for the welding process on each test specimen.
   1.2 Set the Miller Multimatic® 255 welder (Appleton, WI) [42] to DC electrode positive and provide an Ar75/CO$_2$5 shield gas connection to the machine. Use 19.8 *volts* and 405 *in/min* (1030 *cm/min*) wire feed settings for the

TABLE 6.2. *D1.1 Weld leg size specifications [39]*

TEST MATERIAL THICKNESS (IN)	LEG SIZE	SIZE (*IN*)
0.250	minimum	0.125
0.375	minimum	0.1875

0.250 *in* (6.3 *mm*) test pieces and 21.9 *volts* and 410 *in/min* (1040 *cm/min*) wire feed settings for the 0.375 *in* (9.5 *mm*) specimens.

1.3 Use the gas metal arc welding process to weld two A36 mild steel plates together in a T-joint configuration welded on one side of the joint, as in Figure 6.7.

1.4 Visually inspect the weld for excess porosity and undercut. Void the test specimen if such inclusions are visible.

1.5 Cut a 1 *in* cross-section of the test specimen on both ends of the test specimen with a power hack saw.

1.6 Measure the size of the fillet weld on the remaining 6 *in* (15 *cm*) test specimen with a fillet weld gauge, as in Figure 6.8, or with dial calipers, shown in Figure 6.9, for a more precise measurement. The fillet weld size must be uniform for the entire weld length. Photograph and record measurement according to test specimen number.

FIGURE 6.6. A12 test specimen for fillet weld bending experiment.

FIGURE 6.7. Plate configuration for welding pieces into fillet weld bending experiment test sections.

2. Macroetch Test

   2.1 Grind and polish the welded joint area on one of the 1 *in* (2.5 *cm*) cross-sections.

   2.2 Apply phosphoric acid to the welded joint while the joint is still warm from polishing.

   2.3 Allow 2 *min* to let the phosphoric acid etch the weldment.

   2.4 Visually inspect the joint for adequate root fusion and equal fusion into the base metal. Photograph and record the root fusion.

3. Repeat Steps 1.1 –2.4 for 40 test specimens.

   3.1 Number each test specimen for tracking purposes. Assemble the samples for transport to the test site at the Purdue Pankow Laboratory, as shown in Figure 6.10.

4. Data Collection

   4.1 Install the test specimens, oriented as in Figure 6.11, into an MTS Systems (Eden Prairie, MN) Insight electromechanical press, using rapid travel to bring the head into contact with the test piece. Following contact, zero the data logging equipment to eliminate any offsets. Start the real-time data logger to record and plot load and displacement.

   4.2 Apply continuous loading to the test specimen within the press until the load/displacement curve is completed.

   4.3 Remove the test specimen and inspect the edge of the break for root fusion in compliance with the AWS D1.1 visual inspection criterion for a fillet weld break test.

FIGURE 6.8. Measuring fillet weld with manual fillet weld gauge.

FIGURE 6.9. Measuring fillet weld with dial calipers for more precision.

5. Repeat steps 4.1–4.3 for each sample of the various 0.250 *in* (6.3 *mm*) test specimens within each category of size. Then repeat the process for the other set of associated 0.375 *in* (9.5 *mm*) test samples.

All of the test piece fabrication was performed at the Purdue University Agricultural and Biological Engineering ADM Student Achievement Center during the fall of 2022 by Tyler J. McPheron. To achieve minimal deviation between samples, each test specimen was configured and hand-welded with consistent positions and machine settings. Preliminary test T-joints were fabricated to adjust the welding machine for optimal performance. A macroetch test was conducted on the test T-joints to evaluate the root fusion for different machine settings. The final welding machine specifications are shown in Table 6.3, with some uncertainty and adjustment tolerance noted due to building current draw and minor machine error.

FIGURE 6.10. Weld test specimens for bending experiment with Sample A on left and Sample B on right.

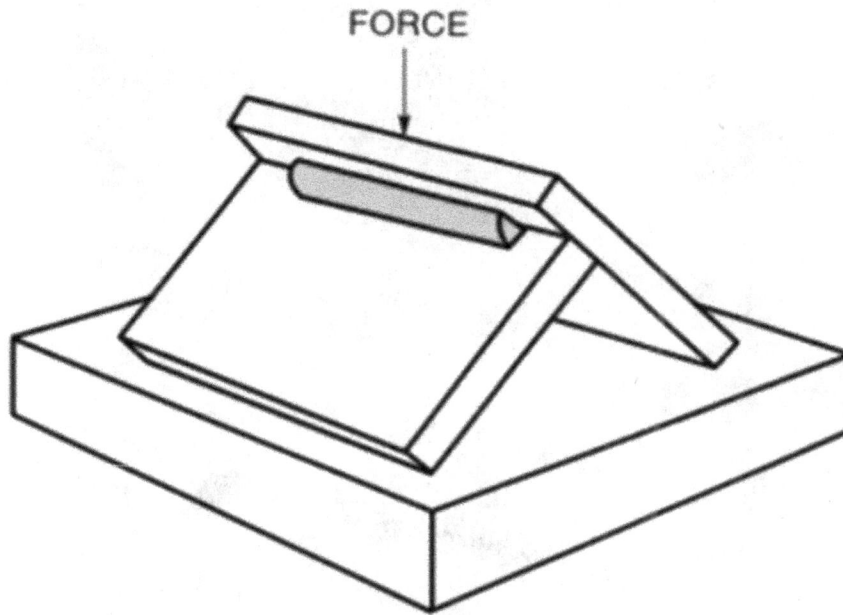

FIGURE 6.11. Fillet weld break test configuration for bending experiment [39].

TABLE 6.3. *Welding equipment specifications for manufacturing welding bend test sections*

WELDER	MILLER MILLERMATIC 252 MIG WELDER	230V
Voltage (V)	19.8 ± 0.5 (For ¼ *in* material)	21.9 ± 0.5 (for 0.375 *in* material)
Wire Feed Speed (in/min)	405 ± 10 (For ¼ *in* material)	410 ± 10 (for 0.375 *in* material)
Shielding Gas	25% Argon, 75% CO_2	—
Filler Metal	E70S Filler Wire (70,000 *psi*)	—
Filler Metal Diameter	0.035 *in*	

Each test specimen was placed in the same position on the welding table, with the ground clamp placed in the same location for all samples. Each test specimen was welded in the 2F horizontal position, with a slight support underneath the specimen for a partial 1F flat welding position. The specimen was configured this way for welder comfort. Figure 6.12 shows a test specimen setup for manufacturing.

3.2 SPECIMEN INSPECTION DETAILS

Table 6.4 details the bending test equipment used in this experiment. An MTS Systems (Eden Prairie, MN) Insight electromechanical press, shown in Figure 6.13, was the primary piece of equipment used to carry out the weld break test experiment. The MTS press had a 300 *kN* capacity rating (67,400 *lb$_f$*). A preliminary finite

FIGURE 6.12. Welded test specimen for bending experiment ready for initial inspection.

TABLE 6.4. *Equipment for data collection and testing welded samples in fillet weld bending experiment*

DESCRIPTION	MFG./MODEL	SPECIFICATION
Electromechanical press	MTS Insight	300 kN standard Length
Load cell	MTS 569331-01	300 kN capacity
Software	Labworks	—

element analysis study was conducted to estimate the amount of force required to break each test specimen or achieve the ultimate strength on the stress/strain curve. The finite element analysis results showed that the joint would fail at approximately 5,000 lb_f (22.2 kN). This preliminary calculation ensured that the MTS press would be sufficient for testing the fillet weld specimens. The MTS press was paired with Labworks (Lehi, UT) software capable of actively plotting the load/displacement curve for each sample. The output values provided a large amount of as-measured data values for each test specimen, thereby meeting the experimental objectives for statistically valid testing.

The AWS has defined criteria for the inspections necessary before and after a fillet weld break test has been conducted. Before destructive testing of the test specimen, visual inspections should be completed by a certified welding inspector. The weld should be reasonably uniform throughout the length of the weld and feature no exterior cracks, excess undercut, or porosity [39]. Assuming that the specimen clears the visual inspections,

FIGURE 6.13. MTS Systems (Eden Prairie, MN) Insight electromechanical press used in fillet weld bending experiment.

the test is considered a pass even if the specimen bends flat upon itself. Many engineers believe that a failed test occurs if the weld fractures. However, if the weld fractures on the centerline of the weld throat, shows complete fusion to the joining base metals, and has no inclusion or porosity larger than 0.094 *in* (2.5 *mm*) in its greatest dimension and if the sum of the greatest dimensions of all inclusions and porosity do not exceed 0.375 *in* (9.5 *mm*) in the 6 *in* (15 *cm*) long specimen, then the test specimen is considered to have passed [39]. It should be known that it is common for the fillet weld to fracture during this test, especially if the base material is 0.500 *in* (12.7 *mm*) or thicker.

A macroetch test was conducted by applying an acidic metal etching chemical to a cross-section of the fillet weld. A variety of chemicals can be used to achieve the etched conditions. In this experiment, a 37% phosphoric acid solution was used to macroetch test each test specimen. Many welds can look uniform and structurally sound on the exterior, but they may have hidden imperfections underneath. The macroetch test is used to ensure that the proper root fusion in the joining material is achieved. The AWS specifies that the weld shall show fusion to the root of the joint but not necessarily in excess amounts [39]. Excessive root fusion can weaken the joint significantly and can cause unacceptable amounts of undercut. This usually indicates too much current on the machine setting or a slow electrode travel speed. The true leg size of the weldment should also be measured at the time of the macroetch test.

3.3 EXPERIMENTAL ASSUMPTIONS

There are certain assumptions necessary to complete the logic chain for the bend testing analysis. The purpose for using the AWS fillet weld break test was to validate the data collected by implementing standard welding procedures. The standardized testing procedure [39] was followed to evaluate the 20 samples of 0.250 *in* (6.3 *mm*) T-joint test specimens and 20 samples of 0.375 *in* (9.5 *mm*) T-joint test specimens and establish a connection to the historically performed experimental weld testing procedures. The overall process validated the data collected and provided the basis for a scientifically binding experiment and conclusions. Additional considerations undertaken were to follow the accepted procedures, measure the necessary parameters outlined by AWS testing criteria, and be willing to exclude any noncompliant test specimens from the results to ensure the validity of the data collected in the experiment. It should be noted during this experiment that a noncertified but experienced welder fabricated the test specimens. In comparison, the 1968 Task Committee of the AWS Structural Welding Committee sent test specimens to a variety of different certified welders in different regions in the United States to ensure uniformity. The ideal scenario would be to program and use a robotic welder to fabricate all the test specimens examined in the experiment, but this was a preliminary study, and the resources for this experimentation were limited.

The purpose for using the AWS fillet weld break test was to authenticate the results by implementing standard welding procedures to collect the data while following a standardized bend testing procedure for the samples of 0.250 *in* (6.3 *mm*) T-joint test specimens and samples of 0.375 *in* (9.3 *mm*) T-joint test specimens. The AWS D1.1 procedure for the fillet weld break test calls for the test specimen to be quenched in water directly after the weld is completed. The test specimens used in this experiment were not quenched after welding to ensure that the full tensile strength of the weld material was considered for analysis. The AWS requires weld quenching for certification tests that allow the weld to break easier. Not quenching the test specimens can alter the way the weld breaks, since the weld filler metal has a higher tensile strength than the base metal. Certified welding inspectors do not measure failure load for the fillet weld break test because most of the focus is on visual inspection and observation. These tests are completed to evaluate the welder's ability to join two materials while maintaining proper weld size and adequate root fusion within the joint. These parameters can be measured more accurately if the weld breaks at the center of the root. Under those circumstances, more attention is directed toward the inspection if the break favors one side or the other. In this experimental investigation, loading at failure was of direct interest.

4. RESULTS AND DISCUSSION

The measured values for the bend break weld tests of the fabricated specimens included load, time, and crush head extension of the press. Each test specimen was fixtured in the MTS machine in the same orientation, matching the AWS fillet weld break test loading criteria. The data recording equipment was then zeroed

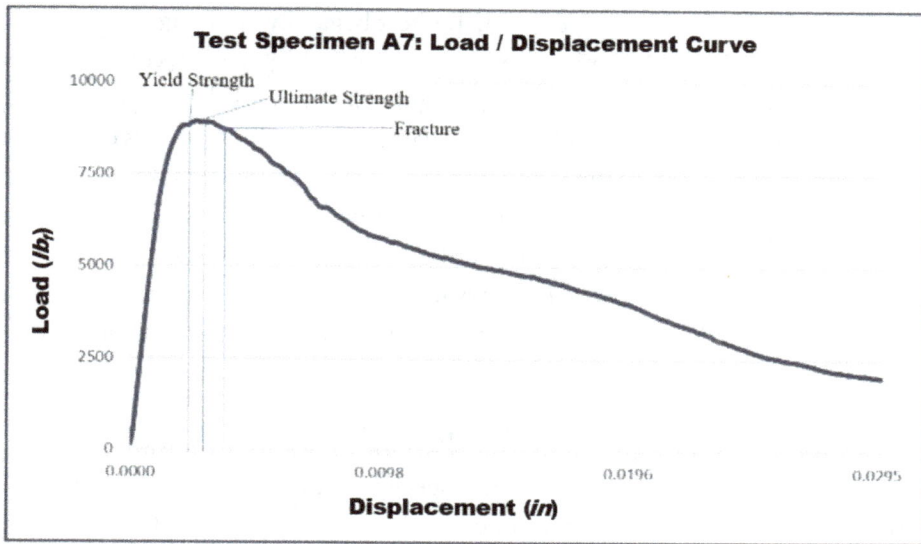

FIGURE 6.14. Stress/strain curve for A7 test specimen in fillet weld bending experiment.

FIGURE 6.15. Test specimen in MTS Systems (Eden Prairie, MN) Insight electromechanical press during fillet weld bending experiment.

using the software program, before beginning the test. The test began by applying a continuous load to the test piece. These tests produced the load/displacement curves for each specimen, containing the maximum amount of load each sample absorbed before fracture [4]. Figure 6.14 presents the results for test specimen A7, which were typical. The highest load was typically reached within the first few seconds of initializing the test. The crushing was stopped once the load declined to 1,000 lb_f (4.4 kN). These steps were repeated for all 0.250 in (6.3 mm) thick and all 0.375 in (9.3 mm) thick test specimens. Figure 6.15 shows a crush test in progress. The remainder of this section will discuss the bend break testing, the root fusion inspections, the statistical analysis of the bend test data, a comparison of the loading results between the two sample sizes, and an initial estimate of the proposed strength reduction design criteria, based on the current experimentation.

4.1 BEND BREAK TESTING

The loading results at failure were recorded for each test specimen. Consistent results were achieved for the 0.250 in (6.3 mm) thick A test specimens, with a mean failure load value of 6,640 lb_f (29,500 N) and a standard deviation of 909 lb_f (4,040 N), resulting in a coefficient of variation of 0.137. This was a good result, considering the amount of potential error that can occur by manually welding each test specimen. It was also interesting to note that the A7 sample featured a slightly smaller weld size than the rest of the samples, yet it failed above the sample average failure load at 7,010 lb_f (31,200 N). This was an initial indication that there can be several variables that affect the strength of a weld rather than just the effects of increasing or decreasing the weld size. It could be that there was a greater root penetration in the weldment, producing a small leg size, since more weld material would be concentrated within the joint in that case.

Table 6.5 shows the aggregate summary results for 0.250 in (6.3 mm) thick A test specimens and the 0.375 in (9.3 mm) B test specimens. Results for the B test specimens were less uniform than the results for the thinner A test specimens. However, for the size of the sample, these results will suffice. The potential error from manually welding and preparing each test sample created a large variability in results. The mean failure load for sample B was 12,400 lb_f (55,160 N), with a standard deviation of 1,600 lb_f (7,120 N), resulting in a coefficient of variation of 0.129. Even though the average size of the weld for B sample was only 12% higher than the average weld size for A samples, a 60% increase in ultimate load capacity was observed for the larger pieces. It is important to clarify that the results presented represent the raw, unfiltered data for the experiment. This data had not yet been cleaned and evaluated for sample failures and outliers that could potentially skew results and therefore was not used for the study's statistical analysis of the experiments.

4.2 BREAK TEST ROOT FUSION INSPECTIONS

The MTS electromechanical press did not typically break the test specimens apart during testing. The load was lifted from each test specimen once it decreased to 1,000 lb_f (4.4 kN). If the test specimen were to break before the end of the load curve, this would indicate a very weak weld with little penetration. Each T-joint was manually separated for further inspection. The main objective for the fillet weld break test was to examine root penetration at the joint and the distribution of weld metal within the base metals. This was done by separating the T-joint and inspecting the break. A clean break directly down the center axis of the weld indicated that there was an equal distribution of weld metal on both joining materials. This also implied that root penetration was symmetric and did not favor one side of the joint or the other. Favoring either leg was an

TABLE 6.5. *Measured data summary statistics for the two thickness treatments with the fillet weld bending experiment*

SAMPLE	MATERIAL THICKNESS (IN)	μ LEG SIZE (IN)	μ THROAT SIZE (IN)	μ FAILURE LOAD (LB$_f$)	FAILURE LOAD STD. DEV. (LB$_f$)
A	0.25	0.247	0.175	6,641	909
B	0.375	0.278	0.196	12,371	1,603

TABLE 6.6. *AWS fillet weld break test acceptance criteria [40]*

THE BROKEN SPECIMEN SHALL PASS IF:

(1) The specimen bends flat upon itself.

(2) The fillet weld, if fractured, has a fracture surface showing complete fusion to the root of the joint.

(3) No inclusions or porosity larger than 3/32 *in* (2.5 *mm*) in its greatest dimension

(4) The sum of the greatest dimensions of all inclusions and porosity shall not exceed 0.375 in (10 *mm*) in the 6 *in* (15 *cm*) long specimen.

indication that electrode angle was either too steep or too shallow in relation to the joint at that point. Table 6.6 presents the overall weld failure points for specimen testing.

Two of the T-joints failed inspection due to excessive undercut. Undercut is an area of concentrated penetration into the base metal, usually at the toe of the weld, creating a visible crevice, further reducing the strength in this area significantly. The two rejected pieces with undercut were likely caused by welding too quickly. All pieces were fabricated by a single welder working under a deadline. The psychology of manufacturing so many test pieces for a time-constrained graduate-level academic experiment certainly could have created these flaws. Figure 6.16 shows one of these deficient weldments. Eleven other T-joints failed inspection due to lack of root penetration. It is important to emphasize that there cannot be any amount of area along the joint that does not fuse into both base metals. Most of the failed T-joints had only 0.125 *in* (3.2 *mm*) to 0.250 *in* (6.3 *mm*) length of effective weld area without proper root fusion, but this is still unacceptable for the bend testing and validation purposes. The cause of the penetration issues was also likely human error, with the torch tip being incorrectly aligned with the test pieces and concentrating filler metal on the bottom plate rather than creating an even distribution having good penetration. Figures 6.17 and 6.18 present examples of failed weldments with insufficient root penetration.

4.3 STATISTICAL ANALYSIS

To analyze the overall results of the experiment, four statistical tests were performed. Each research question was articulated to analyze one of the key objectives of the experiment. These tests, along with their hypothesis, are shown in Table 6.7. There was enough data to do multiple analysis relating to the load and displacement curves produced for each sample. However, the primary focus for this experiment was on the ultimate strength of the weld in failure loading, the effective area of the weld, and the relationship between calculated

FIGURE 6.16. Undercut on failed B$_{II}$ test specimen in fillet weld bending experiment.

FIGURE 6.17. Failed A$_I$ test specimen through lack of fusion into bottom plate during fillet weld bending experiment before testing.

theoretical strength and measured ultimate strength. All statistical tests were performed using R programming language for statistical computing.

4.3.1 NORMALITY OF DATA

To prove the normality of the data samples, a Shapiro-Wilks test was performed for the raw data samples and the cleaned data samples with the failed test specimens removed. A p-value greater than the significance level of $\alpha = 0.05$ was observed for all sample scenarios under Shapiro-Wilk testing. The data with the removed test specimens remained normally distributed, and the exclusion of the failed samples did not change the distribution of the data sets. Results of the normality test are presented in Table 6.8.

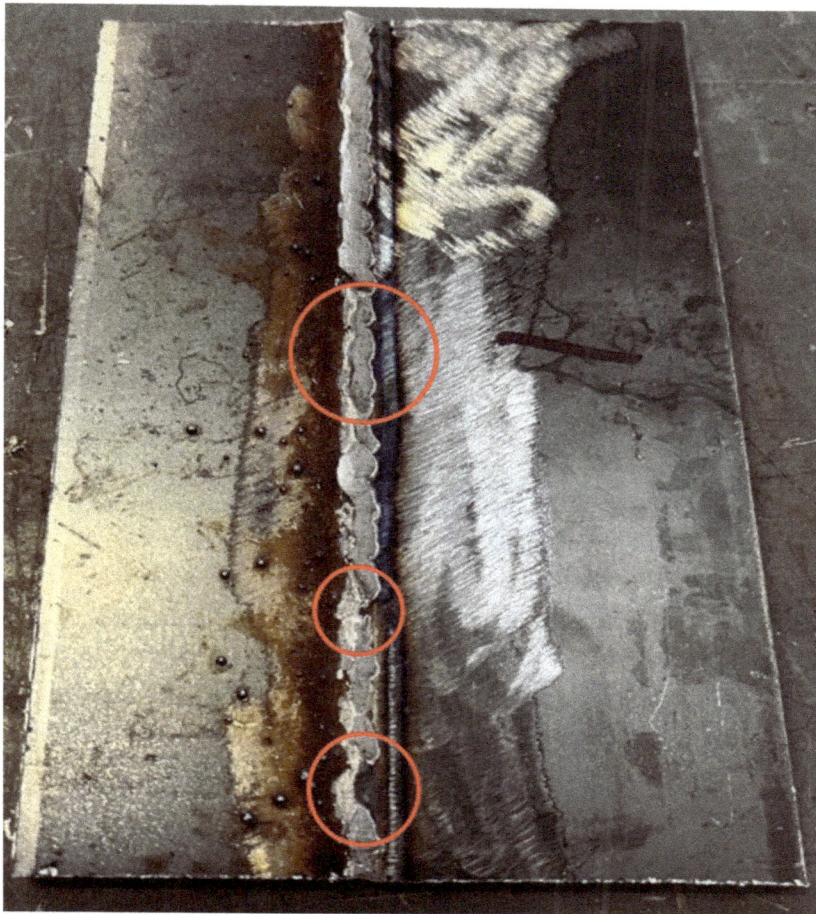

FIGURE 6.18. Failed A1 test specimen through lack of fusion into bottom plate during fillet weld bending experiment after testing.

4.3.2 DATA VERSUS CLEANED DATA

Failed post-fabrication inspections in compliance with AWS D1.1 inspection criteria eliminated 13 of 40 test specimens due to either a lack of root fusion observed after the break or a lack of root fusion observed when performing the macroetch test before the break test. Five of the failed pieces were in Sample A and eight of the failed pieces were in Sample B. This left 15 specimens for Sample A and 12 specimens for Sample B. However, there were some interesting findings that come from comparing the raw data with the clean data for both samples.

It might have been reasonably expected to see a lower failure load from the test samples that were not compliant with the standards, but that was not necessarily the case during this experimentation. Many of the failed test samples failed within a reasonable amount of variance from the cleaned sample mean. Another anomaly in the results included test specimens that inspected quite well but then failed at a lower load than other test specimens that failed inspection. For example, Sample B12 failed at 9,470 lb_f (42,100 N), but it showed adequate root fusion, a uniform weld, and a break that ran down the center axis of the weld. The lowest measured

TABLE 6.7. *Statistical analysis research statement summary for fillet weld bending experiment*

STATISTICAL OVERVIEW
1. Failure Load vs. Effective Weld Area
Is there a difference between failure load (lb_f) of 0.25 *in* T-joints (sample A) and 0.375 *in* T-joints (sample B) due to the difference in required weld sizes for the two base material thicknesses? H_o: Sample A failure load = Sample B failure load H_a: Sample A failure load < Sample B failure load Test: Two-sample t-Test
Are there any significant outliers for failure load (lb_f) due to possible impurities in the weld for each sample of T-joints? H_o: Significant outliers for failure load for both samples = 0 Test: Identify outliers
2. Allowable Stress & Design Factor
Is there a difference between calculated equation factors between Sample A and Sample B? H_o: μ equation factor for Sample A = μ equation factor for Sample B H_a: μ equation factor for Sample A ≠ μ equation factor for Sample B Test: Two sample t-Test
Can an accurate conclusion be found as an acceptable equation factor for use in the design of components featuring welded joints loaded in a cantilevered bending configuration? H_o: μ difference of equation factor between Sample A and Sample B = 0 Test: Comparison of means

TABLE 6.8. *Shapiro-Wilk data normality test results for fillet weld bending experiment data*

SAMPLE A	
Raw data	p-value = 0.727
Cleaned data	p-value = 0.121
SAMPLE B	
Raw data	p-value = 0.434
Cleaned data	p-value = 0.086

failure load from the rejected test specimens in Sample B was measured at 10,230 lb_f (45,500 *N*). Table 6.9 contains statistical summaries for the failed test specimens, comparing them against the passing test specimens.

The passing test specimens characterized the population well, even following the sample size reduction from removing the failed test specimens. The summary statistics were reanalyzed to validate the reliability of the smaller sample sizes after removing the failed test specimens. The normality check using the Shapiro-Wilks test confirmed that the cleaned data was adequately normal to perform the remaining statistical comparisons tests for the experiment.

TABLE 6.9. *Summary statistics for Samples A and B in the fillet weld bending experiment*

	SAMPLE A			SAMPLE B	
FAILURE LOAD	INSPECTION: P/F	LOAD (LB.)	FAILURE LOAD	INSPECTION: P/F	LOAD (LB.)
μ	Pass	6,870	μ	Pass	12,546
μ	Fail	5,956	μ	Fail	12,110
Max	Pass	8,411	Max	Pass	15,360
Max	Fail	7,449	Max	Fail	14,592
Min	Pass	5,930	Min	Pass	9,474
Min	Fail	4,947	Min	Fail	10,231

4.3.3 OUTLIERS FROM DEFECTS AND IMPURITIES

A comparison of the results for the bending experiment of the failed versus passed sample pieces is presented in Table 6.10. There were no significant outliers identified by performing an outlier's test on the raw data for Sample A and Sample B. This analysis considered all 40 T-joint test specimens. A box plot, shown in Figure 6.19, was used to identify any outliers that might have been present in the data. The lack of outlier detection in the raw data inferred that none of the 13 test samples that failed the AWS root fusion inspection were actually statistical outliers. Furthermore, the mean ultimate strength of the samples that failed the fillet weld break test inspections did not deviate significantly from the mean ultimate strength of the test samples that passed the fillet weld break test inspections. Therefore, it can be concluded that minor amounts of nonexistent root fusion, undercut, and impurities do not reveal themselves as significant outliers or have a direct correlation to ultimate strength. However, it may be assumed that multiple imperfections or the lack of root fusion, such as for more than 25% of the effective weld area, would likely cause a notable reduction in ultimate strength.

Despite the lack of statistical outliers in the weld bend test specimens, there were a few interesting occurrences to note. Test sample B12 showed an ultimate strength of 9,470 lb_f (42,100 N). This was well below the mean ultimate strength of 12,550 lb_f (55,800 N) for the samples that passed inspection, yet the B12 sample inspected well, with a uniform break through the center axis of the weld and adequate root fusion at the joint. The premature failure compared to the rest of the samples could have been caused by a subtle variation when fixturing in the MTS machine or by quality issues within the steel material, particularly that the grain direction was not considered when making the samples. Another potential reason for premature failure might include an improper fit-up during fabrication prior to welding. There were no visible imperfections or quality issues with this sample, but its substandard performance was noteworthy. Figures 6.20 and 6.21 show this unique specimen.

4.4 FAILURE LOAD VERSUS EFFECTIVE WELD AREA

For the 0.250 *in* (6.3 *mm*) and 0.375 *in* (9.5 *mm*) plate thicknesses evaluated in the experiment, different weld size standards were specified and created during the fabrication of the samples. An exact measurement of the produced weld is vital to estimating its strength. Figure 6.22 displays some tools for weldment measurement. Figure 6.23 illustrates the relative geometry of the weldment under study, and equations [1] and [2] define the

TABLE 6.10. *Comparison of pieces tested with passing means versus pieces tested with failing means in the fillet weld bending experiment*

SAMPLE A		
	Mean ultimate strength	6,870 lb_f
Mean ultimate strength of failed samples		5,956 lb_f
SAMPLE B		
	Mean ultimate strength	12,546 lb_f
Mean ultimate strength of failed samples		12,110 lb_f

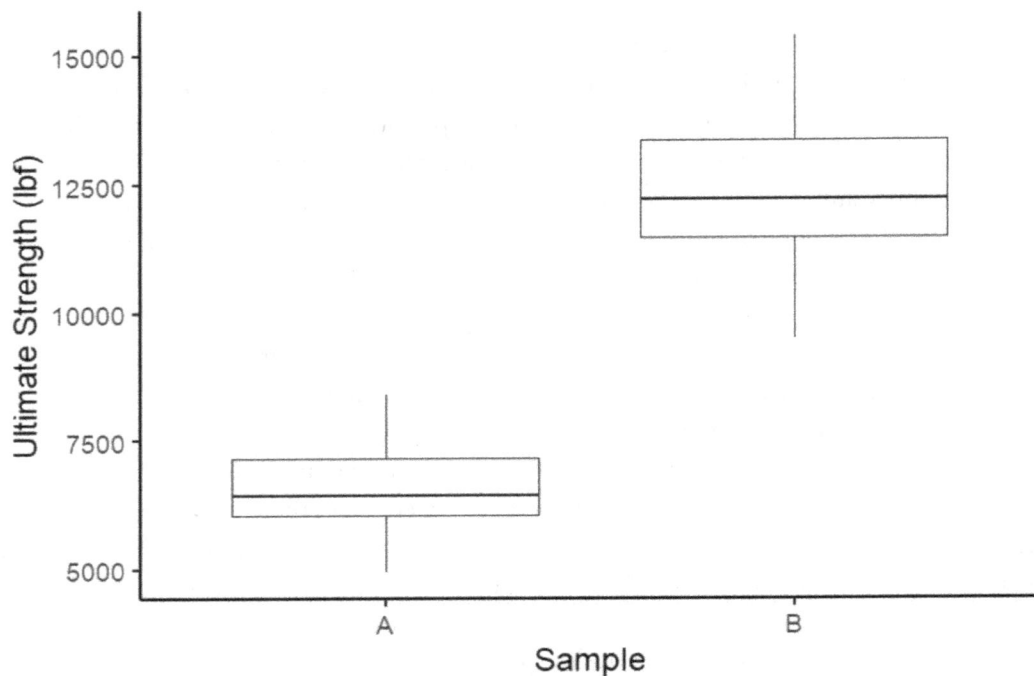

FIGURE 6.19. Uncleaned data outlier detection test for the fillet weld bending experiment.

basic parameters of throat thickness and effective weld area. Table 6.11 shows that the fillet weld leg size can be a minimum of 0.125 *in* (3.2 *mm*) for a fillet weld on 0.250 *in* (6.3 *mm*) material [39]. The fillet weld leg size can be a minimum of 0.188 *in* (4.8 *mm*) for a fillet weld on 0.375 *in* (9.5 *mm*) material [39]. Previous editions of the AWS D1.1 Structural Welding Code specified a maximum leg size for different material thicknesses, but this constraint has been removed in recent revisions of the code. Unfortunately, this revised guideline does not then constrain overwelding. Overwelding can be very costly for manufacturers of welded parts. Overwelding does not increase the strength of the joint and in some cases may noticeably weaken the joint. Welders and

FIGURE 6.20. B12 test specimen from the fillet weld bending experiment.

FIGURE 6.21. Fracture down the center of the weld axis of the B12 test specimen from the fillet weld bending experiment.

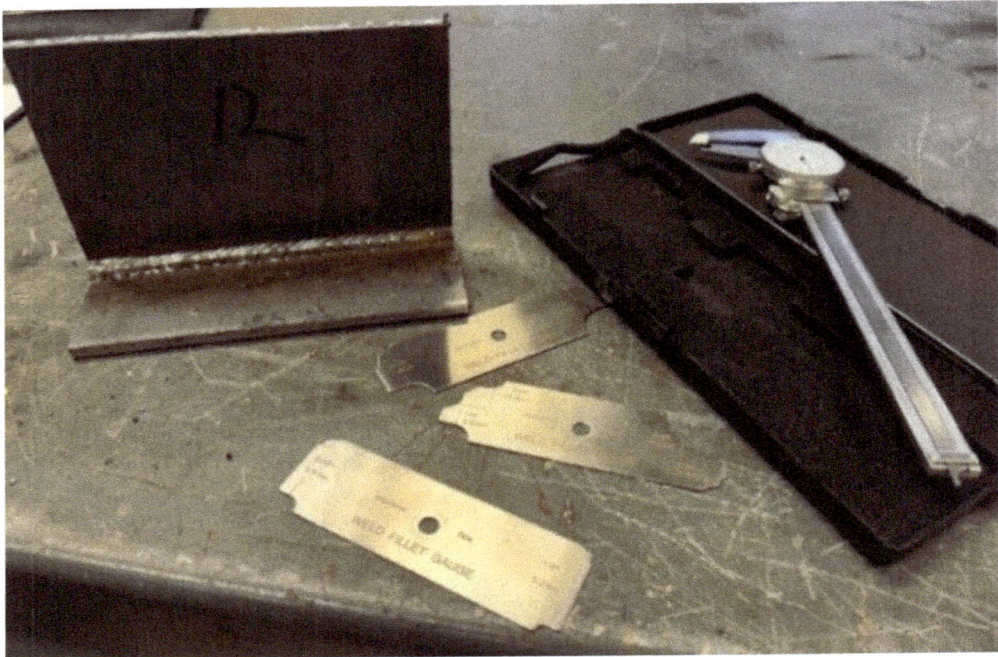

FIGURE 6.22. Fillet weld measurement tools.

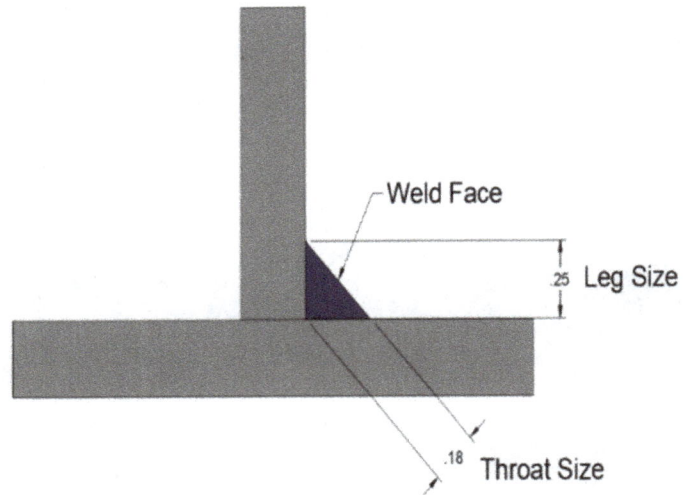

FIGURE 6.23. Cross-section of fillet welded T-joint.

TABLE 6.11. *Welding leg sizes for Samples A and B for the fillet weld bending experiment*

SAMPLE A		
	Min weld leg size	0.1875 in
	μ Weld leg size	0.246 in
	μ Effective weld area	1.044 in^2
SAMPLE B		
	Min weld leg size	0.250 in
	μ Weld leg size	0.278 in
	μ Effective weld area	1.187 in^2

engineers use a common rule of thumb for a fillet weld leg size that is 75% of the base metal thickness. This maintains adequate strength while minimizing overwelding [43]. In this experiment, all fillet weld leg sizes were measured with dial calipers and fillet weld gauges to accurately quantify the leg size for each test specimen. Fillet weld gauges are good tools for taking quick measurements but have limited accuracy. A measurement of 0.001 *in* (0.0025 *mm*) can be obtained by using dial calipers. Obtaining the leg length and the overall length of the weld for each test specimen provided the necessary measurements to calculate the effective weld area for each test specimen under Equations 1 and 2.

$$W_t = (\cos(45°)) * L_S \qquad [1]$$

$$A = W_t * L_w \qquad [2]$$

where W_t = theoretical throat thickness (*in*), L_s = leg size (*in*). A = effective weld area (*in²*), and L_w = weld length (*in*).

4.4.1 FAILURE LOAD VERSUS EFFECTIVE WELD AREA ANALYSIS

The mean failure load for the thicker 0.375 *in* (9.5 *mm*) test specimens in Sample B was approximately twice the mean failure load observed for the 0.250 *in* (6.3 *mm*) test specimens in Sample A. Due to these significantly different results, the hypothesis that failure load was equal, despite the larger weld size and thicker base materials, would be rejected. Table 6.12 provides the details of these statistical results. It is difficult to tell from these results if the plate thickness or the effective weld area influenced the strength of the weld more than the other. Figure 6.21 showed the simplified cross-section of a fillet welded T-joint. However, this is not the realistic shape of an appropriate weld. If the weld was completed properly, the theoretical size of the weld throat should be a close approximation to the actual weld size, but it is typically less than the actual weld size. Additionally, the increased weld size in Sample B for the larger plate thickness compared to Sample A was likely a contributing factor to the Sample B average failure load being twice the ultimate average load capacity of Sample A. Larger base metal thicknesses provided greater structural area and support for the joint when bend tested in the fixturing configuration used for this experiment.

The difference between the theoretical throat size in Figure 6.21 and the larger actual throat size shown in Figure 6.24 is notable. The root depth and penetration into the base materials is greater in the thicker

TABLE 6.12. *Statistical summary of results for failure load versus effective weld area for fillet weld bending experiment*

TWO-SAMPLE T-TEST

SAMPLE	MEAN (LB_F)	STD DEV (LB_F)	P-VALUE	T-VALUE	DF	MIN (LB_F)	MAX (LB_F)
A	6,870	774	p<0.001	−11.81	25	5,930	8,411
B	12,546	1,654	p<0.001	−11.81	25	9,474	15,360

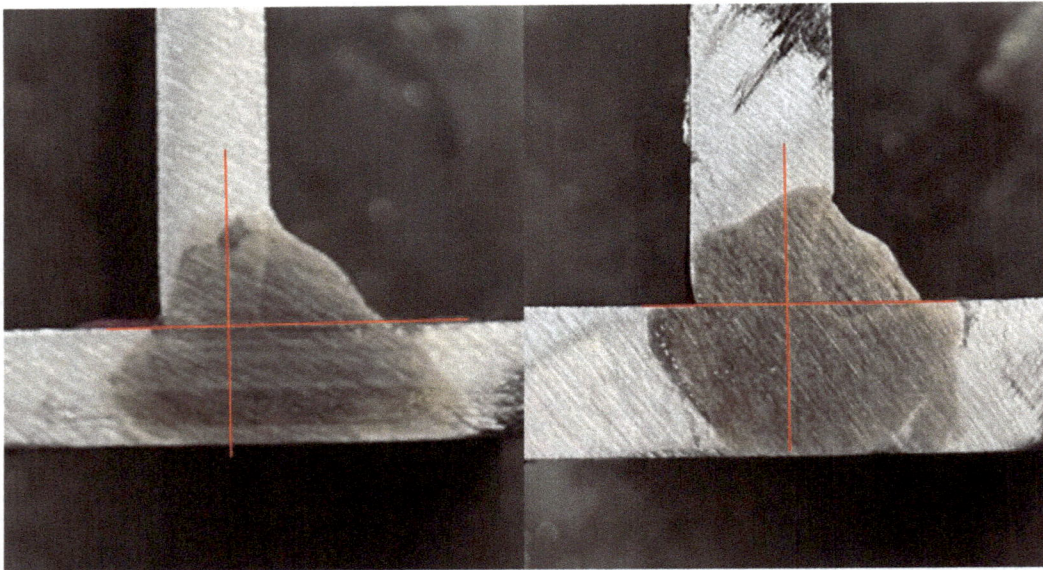

FIGURE 6.24. Sample A macroetch test versus Sample B macroetch test for fillet weld bending experiment.

specimens. The additional material thickness in Sample B is a secondary contributor to the increase in strength measured in the experiment. The mean difference in effective weld area between Sample A and Sample B is 0.143 *in²* (0.923 *cm²*), with a weld leg size difference of 0.032 *in* (0.810 *cm*). With just over 0.0312 *in* (0.79 *cm*) difference between the two samples, this variance might seem negligible. However, the small difference showed profound results when comparing the calculated theoretical load capacity for both weld sizes.

4.4.2 THEORETICAL STRENGTH FOR TRANSVERSE LOADING

Equation [3] provides the means to determine weld strength. By using the previous equations, with a 0.0312 *in* (0.79 *cm*) larger leg size, a 13% increase in theoretical strength was observed. The additional 37% increase in strength seen in the experiment can be credited to the bigger 0.375 *in* (9.5 *mm*) plate thickness. Larger weld sizes have an indirect effect on the strength of a welded connection when subject to cantilevered bending. However, because of the loosely defined standards provided by the AWS for weld sizes dependent on plate thickness, there are still several unknowns as to the effect of plate thickness on strength of the welded connection. A better experiment for this test would be to compare two samples with plate thicknesses inside of the same base metal thickness interval defined by the AWS for one singular weld size. For example, according to AWS D1.1, the minimum leg size required for welding both 0.125 in (3.2 mm) and 0.188 in (4.8 mm) thick base metal is 0.125 in (3.2 mm).

$$F_t = \sigma_t * A \qquad [3]$$

where F_t = weld strength (lb_f), σ_t = ultimate tensile strength of weld filler (psi), and A = effective area of the weldment (in^2).

Testing two samples with these specifications would likely lead to a better indication of the influence that plate thickness has on joint strength. This experiment used 0.250 in (6.3 mm) and 0.375 in (9.5 mm) thick base metals, which require two different minimum weld leg sizes. Therefore, there are two confounding factors potentially influencing the strength of the weld [6]. Calculated weld strengths for the two plate thicknesses are dependent on the measured weld area and projected strength of the resolidified melt volume. Uncertainty in the outcome could come from errors in measurement and variance in projected strength. The latter is likely to be a larger factor, since the strength of the resolidified pool is dependent on multiple factors that could result from the heating and cooling processes. Using the 6 in (15.2 cm) sample length and average measured leg size of the specimens from Table 6.11, the estimated effective area for Samples A and B is 1.044 in^2 (6.735 cm^2) and 1.187 in^2 (7.658 cm^2), respectively. Specimen strength is calculated as

Sample A: 73,080 lb_f = 70,000 psi * 1.044 in^2 = 325 kN
Sample B: 83,090 lb_f = 70,000 psi * 1.187 in^2 = 370 kN

4.5 ALLOWABLE STRESSES AND DESIGN FACTORS

The allowable stress for fillet welds loaded perpendicular to the weld axis is the same as the ultimate tensile strength of the filler metal. The weld filler metal is the material that gets melted into two base materials to join them together. Filler metals are used in all welding processes and must have better or equal mechanical and chemical properties than the base material being welded [44]. Most filler metals used for welding mild low-carbon steel feature an ultimate tensile strength between 60 ksi (414 MPa) and 70 ksi (483 MPa). However, as previously mentioned, AWS D1.1 Clause 4 requires a 70% reduction in the allowable stress for evaluating fillet welds loaded in the shear direction [39]. This allowable stress reduction metric yields safety factors of 2.2 for parallel forces to the weld axis to 4.6 for forces normal to the weld axis [45]. Figures 85 and 86 provide a comparison between transverse and shear loading configurations. To reduce the allowable stress by 70%, the ultimate tensile stress of the filler metal used to produce the weld being evaluated is multiplied by 0.30, as shown in equation [4]. This value was derived from the experiments outlined in Higgins and Preece [37].

$$F_t = \sigma_t * 0.30) * A \qquad [4]$$

where F_t = weld strength (lb_f), σ_t = ultimate tensile strength of weld filler (psi), and A = effective area of the weldment (in^2).

4.5.1 ALLOWABLE STRESS IN T-JOINT WELDING TESTS

The factor for the allowable stress used for estimating the strength of a fillet weld in a bending configuration can be derived by rearranging equation [4] into equation [5]. This equation uses the failure load recorded in the experiment for each test specimen, along with its corresponding measured effective weld area. Table 6.13 provides the results of this experiment in the form of a similar strength reduction stress factor. Using this methodology, the 0.250 *in* (6.3 *mm*) specimens produced a reduction factor of 0.10, and the 0.375 *in* (9.5 *mm*) samples yielded a factor 0.15. As expected, both values are significantly lower than the full-strength transverse loading factor. The cantilevered loading produced a stress reduction factor 50% smaller than that for shear loading.

$$E_f = P_{ultimate} / (\sigma_t {}^* A) \tag{5}$$

where E_f = allowable stress reduction factor, $P_{ultimate}$ = failure load from experiment [ultimate strength] (*lb*), σ_t = ultimate tensile strength of weld filler (*psi*), and A = effective area of the weldment (*in²*).

4.5.2 ALLOWABLE STRESS EXPERIMENT STATISTICS

Results for a consistent allowable stress reduction factor between the two samples were not equal. For this experiment, it can be concluded that hypothesis of equal allowable stress factors calculated from the data should be rejected. A reduction of 90% and 85% for the allowable stress at the welded connection is shown by the data. This is significantly different than the 70% reduction in shear, with forces acting on the weld axis in parallel. It is clear that the bending scenario illustrated in this experiment does not compare well to strength values from welds loaded in shear and transverse conditions. To use this factor in a structural design with similar boundary conditions and geometry, the ultimate strength of the weld would be subject to a considerable reduction for the design engineer. Table 6.14 shows the statistical analysis summary for the weldment bending experiment, clearly indicating a difference in strength reduction between the two thicknesses.

4.6 DISCUSSION

Considering all the test specimens from both samples, no significant outliers in loading strength performance were discovered from any imperfections or impurities within the welds. However, none of the 13 test specimens that failed the post-break test inspection were outliers in the load data. Since the measurements were consistent between the passing test specimens and the failed test specimens, the same conclusions would have likely been drawn from the statistical tests using either the cleaned data or the raw data. Alternatively, welds

TABLE 6.13. *Calculations with averaged test measurements for fillet weld bending experiment*

SAMPLE	FAILURE LOAD (LB)	STRESS (PSI)	TENSILE STRENGTH (PSI)	EFFECTIVE WELD AREA (IN²)	STRESS FACTOR
A	6,870	8,640	70,000	1.044	0.1
B	12,546	15,970	70,000	1.187	0.15

TABLE 6.14. *Allowable stress factor summary statistics for fillet weld bending experiment*

TWO-SAMPLE t-TEST							
SAMPLE	MEAN	STD DEV	P-VALUE	T-VALUE	DF	MIN	MAX
A	0.10	0.01	p<0.001	−7.97	25	0.08	0.13
B	0.15	0.02	P<0.001	−7.97	25	0.11	0.19

that lacked large amounts of root fusion or had large slag inclusions and porosity would have likely demonstrated underperforming results when compared to results achieved by specimens with sufficient welds.

Failure loads measured for each sample set yielded drastically different values, with consistent results measured from each specimen. The T-joint test specimens constructed with thicker 0.375 *in* (9.3 *mm*) A36 steel material showed an average failure load of twice the average failure load of the test specimens constructed with 0.250 *in* (6.3 *mm*) material. The difficulty was in telling how much of the difference in effective weld area between the two samples and the difference in base metal thickness influenced the ultimate load capacity. Due to the loading configuration, there could have been additional structural stability derived from the thicker materials, allowing the welded connection to withstand greater loads. The larger weld size, if properly constructed, did increase the amount of fusion into the base metals and would certainly have increased the strength of the connection. However, it is difficult to quantify the influence of these parameters between the two different plate thickness samples, requiring different weld sizes to comply with AWS D1.1 weld size specifications.

5. CONCLUSION

Two sizes of weld bending test specimens were fabricated to evaluate strength in bending. The number of test piece rejections based on preliminary inspections was 13. While this was a fairly large number, it did not adversely affect the statistical analysis in this study, but it did demonstrate the difficulty associated with producing uniform test pieces for welding experimentation and confirmed the wisdom of the original researchers in having multiple welders produce test pieces for examination. Since 27 out of 39 test specimens passed inspection after destructive testing, this proved that visually appealing welds do not guarantee structural integrity. This experiment showed that welds with uniform and aesthetic appearance may not possess all the critical requirements for a structurally sound weld. In general, the thicker test pieces did produce higher loadings than the thinner pieces, and they did so by an amount larger than the simple addition of more material might suggest. In this experiment, ultimate tensile strength reductions of 90% for Sample A and 85% for Sample B were determined by using the ultimate load capacity measured in the experiment and the measured effective weld area of each test specimen. It can be concluded that welds loaded in cantilevered bending configuration experience more stresses and are subject to failure at loading prematurely compared to that of shear load configurations and certainly compared with transversely loaded fillet welds, which use the total rating of weld filler metal tensile strength for evaluating load capacity of a weld. It is clear why this type of load configuration is not recommended for use in design. When forced into use in a repair situation, the welder must use extreme care to produce good welds and include far longer weld runs than would normally be needed to handle similar loads in other configurations.

FIGURE 6.25. Double fillet welded T-joint as an indicator piece for the fillet weld bending experiment.

Additionally, the strength of a fillet welded connection increases significantly when it is welded on both sides of the joint. The tests in this experiment were all configured with a single fillet weld on one side of the connection to follow the AWS test. A T-joint welded properly on both sides of the joint will likely never fail in this testing configuration. The double-sided T-joint test piece shown in Figure 6.25 was also crushed in the MTS 150 *ton* (136 *mton*) hydraulic press to illustrate the increased strength that the double fillet weld provides. When applying the maximum load to the double fillet welded test piece, there was permanent deformation, but there was no fracture at the weld. The amount of weld mass at the joint provided significant structural support to the base materials.

The findings from this work can be summarized in three main points:

- Aesthetic welds are not always good welds. Good design, best procedural practices, and quality assurance are necessary to manufacture a quality welded product.
- Appropriate material selection is important for proper load carrying capacity. Thicker materials and deeper welds provide more load carrying area.
- Further testing of the welded joint cantilever bending scenario seems justified. In the meantime, the results from this study would seem to indicate that extreme caution should be used in any repair contemplating a single-sided weld, and a double-sided weld should be used in design and, where possible, within repairs.

For further experimentation and research, it is recommended that the human variation within the welding process and construction of each sample be eliminated by using robotic welders, similar to those currently used in many industrial manufacturing processes. A numerical test piece fabrication methodology would likely result in less imperfections and far more consistent test data. The designers of the 1968 working stress experiment aimed to minimize the human error factor for their samples by having many welders from different locations in the United States weld a specific number of test specimens, which were then sent back to the researchers for testing. Increasing the number of total samples should also be recommended for improving test results. A high sample number allows for a valid statistical analysis, even if some of samples fail post-destructive testing inspections. A similar experiment using two material thicknesses in the same recommended weld size range should provide adequate testing to evaluate the effects of weld size on ultimate strength, which were confounded in this study. In the past, inferior welds were being made to look like adequate welds, and this is the reason for the development of the standardized codes for weld sizes. However, these codes were developed to eliminate bad habits and poor welding practices. There are also many examples where these specifications may not be applicable to the design. It is up to the design engineers to consider all possible scenarios and parameters associated with unique product assembly problems [45]. An experiment that seeks to narrow the required fillet weld size specification for this bending condition would provide design engineers and welders with a better guide for deciding what fillet weld size should be used for this unusual application. A study that yielded a more defined size specification would ensure more consistent welds that provide adequate structural integrity and have a greater economic impact, especially in the repair of manufactured components.

6. ACKNOWLEDGMENTS

The authors would like to acknowledge and thank Mr. Jeffrey Lynch of Purdue University Research Machine Services and Mr. Scott Brand of Purdue University Agricultural and Biological Engineering. Their assistance with this project has been essential to its completion. Additionally, Mr. Robert Hershberger of the Purdue University Civil Engineering Pankow Laboratory is thanked for his guidance in the testing of the sample pieces. Mr. Robert M. Stwalley IV is graciously thanked for his technical illustration support. Dr. Carol S. Stwalley is acknowledged for her statistical and editorial assistance on the article. This research did not receive any specific grant funding from any agencies in the public, commercial, or not-for-profit sectors. However, the assistance of the Purdue University Department of Agricultural and Biological Engineering is gratefully acknowledged for its support over the years with graduate teaching assistance positions and faculty salaries. The mention of trade names or commercial products in this article is solely for the purpose of providing specific technical information and does not imply recommendation or endorsement by Purdue University. The findings and conclusions in this publication are those of the authors and should not be construed to represent any official Purdue University determination or policy. Purdue University is an equal opportunity/equal access organization.

7. EXPERIMENTAL WELD BEND STRENGTH QUESTIONS

1. Describe the effect of stress concentrations near welds and how this might negatively influence a cantilever loaded welded joint.
2. If no out of the ordinary failures had been reported using the 1930s era guidelines, what was the rationale for reducing the effective strength of the shear loaded joint configuration in the 1960s?
3. What expansions and changes to the current experimental program would be needed to extend the general applicability of the results for a broader recommendation?
4. Given the similar coefficient of variability for the bend test results between the two base material sizes, explain the significance of the increase in strength of the larger test pieces.
5. What factors are present within a cantilever loaded single-sided welded joint that decrease its load carrying capability?

8. REFERENCES

[1] R. Budynas and K. Nisbett, *Shigley's Mechanical Engineering Design*, 11th ed. McGraw Hill, 2020.

[2] T. H. Brown Jr., *Marks' Calculations for Machine Design*. McGraw Hill, 2005.

[3] A. D. Deutschman, W. J. Michels, and C. E. Wilson, *Machine Design: Theory and Practice*. Macmillan, 1975.

[4] G. W. Krutz, J. K. Schueller, and P. W. Claar II, "Machine Design for Mobile and Industrial Applications," Society of Automotive Engineers, Inc, 1994.

[5] T. J. McPheron and R. M. Stwalley III, "Engineering Challenges Associated with Welding Filed Repairs," in *Welding Principles and Application*. IntechOpen Limited, 2022.

[6] T. J. McPheron and R. M. Stwalley III, "Fillet Weld Strength Analysis for Canilever Loading," ASABE 2023 AIM-Omaha Conference, St. Joesph, Michigan, 2023.

[7] I. Gomez, A. Kanvinde, Y. K. Kwan, and G. Grondin, "Strength and Ductility of Welded Joints Subjected to Out-of-Plane Bending," 2008. [Online]. Available: https://www.aisc.org/globalassets/aisc/research-library/strength-and-ductility-of-welded-joints-subjected-to-out-of-plane-bending.pdf [Last accessed May 25, 2024].

[8] J. L. Dawe and G. L. Kulak, "Welded Connections under Combined Shear and Moment," *Journal of the Structural Division*, vol. 100, no. 4, pp. 727–741, 1974.

[9] D. K. Dwivedi, *Fundamentals of Metal Joining*. Springer, 2022.

[10] H. B. Cary, *Modern Welding Technology*. Prentice Hall, 2002.

[11] Lincoln Electric Company, *The Procedure Handbook of Arc Welding*. Lincoln Electric Company, 2000.

[12] S.-H. Kim, C.-H. Lee, I. Ryu, and S. Park, "Experimental Investigation of Cold Formed High Strength Steel Tubular Joints Subjected to Moment Loading," *Advances in Structural Engineering*, vol. 26, no. 12, pp. 2172–2198, 2023.

[13] A. H. N. Ateyah, "Bending Stresses Calculation in Welded Joint," *International Journal of Engineering Research and Applications*, vol. 12, no. 1, pp. 1–6, 2022.

[14] A. Santacruz and O. Millelsen, "Numerical Stress Analysis of Tubular Joints," Third Conference of Computational Methods & Ocean Technology, Stavanger, Bristol, UK, 2021.

[15] T. Bjork, A. Ahola, and T. Skriko, "On the Distortion and Warping of Cantilever Beams with Hollow Section," *Welding in the World*, vol. 64, pp. 1269–1278, 2020.

[16] F. I. Islamovic, P. Muratovic, D. Gaco, and F. Kulenovic, "Bend Testing of the Welded Joints," 7th International Scientific Conference on Production Engineering (RIM 2009), Cairo, Bihac, Bosnia and Herzegovina, 2009.

[17] W. Qiu-dong, W. Yang, F. Zhong-qiu, and W. Yi-xun, "Parametric Study on Fatigue Failure Modes of the Rib-Deck Weld under Out-of-Plane Bending Loading," *Iranian Journal of Science and Technology, Transactions of Civil Engineering*, vol. 47, pp. 2625–2637, 2023.

[18] Y. Banno and K. Kinoshita, "Experimental Investigation of Fatigue Strength of Out-of-Plane Gusset Welded Joints under Variable Amplitude Plate Bending Loading in Long Life Region," *Welding in the World*, vol. 66, pp. 1883–1896, 2022.

[19] G. Prayogo, M. A. Budiyanto, and M. Perkasa, "Analysis of Bending Mechanical Performance of Welding Joints with the Addition of Diamond and Circular Plates," *Indonesian Journal of Engineering and Science*, vol. 4, no. 1, pp. 57–73, 2023.

[20] N. A. Husaini, J. K. Hamza, and S. E. Sofyan, "Effects of Welding on the Change of Microstructure and Mechanical Properties of Low Carbon Steel," *Material Science and Engineering*, vol. 523, p. 012065, 2019.

[21] K. Hectors and W. De Waele, "Influence of Weld Geometry on Stress Concentration Factor Distribution in Tubular Joints," *Journal of Construction Steel Research*, vol. 176, p. 106376, 2021.

[22] E. Chouha, S. E. Jalal, Z. El Maskaoui, and A. Chouaf, "Concentrated Stress Location Areas for Welded Tubular T-joint under Deflected Bending Load," *MATEC Web of Conferences*, vol. 286, p. 02004, 2019.

[23] C. M. Mayr and K. Rother, "Improved Stress Concentration Factors for Circular Shafts for Uniaxial and Combined Loading," *Materials Testing*, vol. 61, no. 3, pp. 193–203, 2019.

[24] W. D. Pilkey and D. F. Pilkey, *Peterson's Stress Concentration Factors*, Wiley, 2008.

[25] L. W. Zachary and C. P. Burger, "Stress Concentrations in Double Welded Partial Joint Penetration Butt Welds," *Welsing Research Supplement*, vol. 55, no. 3, pp. 77–82, 1976.

[26] K. Hectors and W. De Waele, "Cumulative Damage and Life Prediction Models for High-Cycle Fatigue of Metals: A Review," *Metals*, vol. 204, no. 11, 2021.

[27] A. Hobbacher, *International Institute of Welding: Recommendations for Fatigue Design of Welded Joints and Components*, Springer International Publishing, 2016.

[28] A. Hobbacher and M. Kassner, "On Relation between Fatigue Properties of Welded Joints, Quality Criteria and Groups in ISO 5817," *Welding in the World*, vol. 56, pp. 153–169, 2012.

[29] B. Jonsson, J. Samuelsson, and G. B. Marquis, "Development of Weld Quality Criteria Based on Fatigue Performance," *Welding in the World*, vol. 55, pp. 79–88, 2011.

[30] S. I. Talabi, O. B. Owolabi, J. A. Adebisi, and T. Yahaya, "Effect of Welding Variables on Mechanical Properties of Low Carbon Steel Welded Joint," *Advances in Production Engineering & Management*, vol. 9, no. 4, pp. 181–186, 2014.

[31] S. P. Tewari, A. Gupta, and J. Prakash, "Effect of Welding Parameters on the Weldability of Material," *International Journal of Engineering Science and Technology*, vol. 2, no. 4, pp. 512–516, 2010.

[32] T. R. Mupoperi and M. Pita, "An Investigation of the Effect of Welding Current on the Mechanical Properties of Mild Steel Joints When Using Arc Welding," IEEE 13th International Conference on Mechanical and Intelligent Manufacturing Technologies, Cape Town, Piscataway, NJ, 2022.

[33] D. Pathak, R. P. Singh, S. Gaur, and V. Balu, "Experimental Investigation of Effects of Welding Current and Electrode Angle on Tensile Strength of Shielded Metal Arc Welded Low Carbon Steel Plates," *Materials Today: Proceedings*, vol. 26, pp. 929–931, 2020.

[34] S. Y. Merchant, "Investigation on Effect of Welding Current on Welding Speed and Hardness of Haz and Weld Metal of Mild Steel," *International Journal of Research in Engineering and Technology*, vol. 4, no. 3, pp. 44–48, 2015.

[35] R. Schiller, M. Oswald, J. Neuhausler, K. Rother, and I. Engelhardt, "Fatigue Strength of Partial Penetration Butt Welds of Mild Steel," *Welding in the World*, vol. 66, pp. 2563–2584, 2022.

[36] Y. Tobe and F. V. Lawrence Jr., "Effect of Inadequate Joint Penetration on Fatigue Resistance of High-Strength Structural Steel Welds," *Supplement to the Welding Journal*, vol. 9, pp. 259–266, 1977.

[37] T. R. Higgins and F. R. Preece, "Proposed Working Stresses for Fillet Welds in Building Construction," *Engineering Journal, American Institute of Steel Construction*, vol. 6, pp. 16–20, 1969.

[38] American Bureau of Welding, "Report of the Structural Steel Welding Committee," American Welding Society, Miami, Florida, 1931.

[39] American Welding Society, "AWS D1.1 Structural Welding Code: Steel," 2020.

[40] American Welding Society, "AWS D1.1 Structural Welding Code: Steel," 2000.

[41] American Welding Society, "AWS B4.0:2016 Standard Methods for Mechanical Testing of Welds," 2016.

[42] Miller Electric Manufacturing, LLC, "Multimatic® 255 Multiprocess Welder w/ EZ-Latch™ Dual Cylinder Running Gear & TIG Kit—208-575V," [Online]. Available: https://www.millerwelds.com/equipment/welders/mig-gmaw/multimatic-255-multiprocess-welder-m30175. [Last accessed May 15, 2024].

[43] Canadian Welding Bureau, "Electrode Certifications," 2023. [Online]. Available: https://www.cwbgroup.org/safety/industry/electrode-certification. [Last accessed May 1, 2023].

[44] Z. Barsoum and M. Khurshid, "Ultimate Strength Capacity of Welded Joints in High Strength Steels," *Procedia Structural Integrity*, vol. 5, pp. 1401–1408, 2017.

[45] Structural Engineering University, "Welding Myths Debunked: Maximum Fillet Weld Thickness," 2017. [Online]. Available: https://learnwithseu.com/welding-myths-debunked-maximum-fillet-weld-thickness/. [Last accessed May 1, 2023].

CONCLUSION

Modern design challenges in the agricultural engineering industry are complex. Many elements of today's challenges are overarching "big picture" issues that do not have general solutions but instead must simply be mitigated. Soil compaction is a classic example of this kind of problem. It is impossible to grow a crop without traveling across fields, yet doing so damages the soil media for the current crop and future crops. Engineers must do what they can to minimize the impact of equipment on the soil, and farm managers must carefully plan operations with the same goals in mind. The collection of data in mobile equipment and the execution of welded field repairs fall into the same complicated realm, as numerous other types of modern design challenges also do. The key in all situations involving complex issues is thorough study of the various factors involved and a results-based approach that focuses on achieving the fundamental objectives of the design while minimizing the influence of negative factors that reduce the effectiveness of the proposed solutions. This approach to design requires a wholistic knowledge of all factors influencing the problem and an open-minded mental attitude willing to look beyond the traditional methodologies generally used within the specific problem area. To address these complex issues, it is critical that engineers not lock themselves into a particular solution too early in the development process. It is important that design teams examine a wide variety of diverse and unusual approaches early in the design process in order to be able to extract the unique and innovative elements from those concepts. Novel and original solutions to complex problems require a synthesis of numerous inventive ideas, and no single individual can possibly bring all of the worthwhile concepts to the table. To succeed in tackling large, all-encompassing problems, one must become well versed in all the background information, be open-minded to novel ideas, and stay persistently focused on achieving a solution. If the problem was easy, it would have already been solved.

COMBINED REFERENCES FROM THE INCLUDED WORKS

Abu-Hamdeh, N. H., Carpenter, T. G., Wood, R. K., & Holmes, R. G. (1995). *Soil compaction of four-wheel drive and tracked tractors under various draft loads.* International Off-Highway & Powerplant Congress & Exposition—Milwaukee. Warrendale, PA: SAE International. https://doi.org/10.4271/952098.

Adamchuk, V. I., Skotnikov, A. V., Speichinger, J. D., & Kocher, M. F. (2004). Development of an instrumented deep-tillage implement for sensing of soil mechanical resistance. *Transactions of the ASAE, 47*(6), 1913–1919. https://doi.org/10.13031/2013.17798.

Al-Aani, F. S. (2019). CAN bus technology for agricultural machine management. Iowa State University Capstones, Theses, and Dissertations. Retrieved May 15, 2021, from https://lib.dr.iastate.edu/etd/17635

Alban, D. H., Host, G. E., Elioff, J. D., & Shadis, D. A. (1994). *Soil and vegetation response to soil compaction and forest floor removal after aspen harvesting.* United States Government, Department of Agriculture. Washington DC: USDA Forest Service. https://doi.org/10.2737/NC-RP-315.

Albiero, D., Garcia, A. P., Umezu, C. K., & de Paulo, R. L. (2021). *Swarm robots in agriculture.* CoRR, abs/2103.06732. https://doi.org/10.48011/asba.v2i1.1144.

American Bureau of Welding. (1931). *Report of the structural steel welding committee.* American Welding Society.

American Society for Testing and Materials. (1996). *About Us—ASTM.* Retrieved December 18, 2021, from ASTM International. https://www.astm.org/ABOUT/overview.html

American Society for Testing Materials. (2021). *Standard test methods for cone penetration of lubricating grease.* Retrieved December 27, 2023, from American Society for Testing Materials. https://www.astm.org/d0217-21a.html

American Welding Society. (2000). *AWS D1.1 structural welding Code—Steel.* American Welding Society.

American Welding Society. (2016). *AWS B4.0:2016 standard methods for mechanical testing of welds.* American Welding Society. Retrieved May 15, 2024, from https://pubs.aws.org/Download_PDFS/B4.0-2016-PV.pdf

American Welding Society. (2017). *Specifications for underwater welding (AWS D3.6M).*

American Welding Society. (2020). *AWS D1.1 structural welding code—Steel.* Miami: American Welding Society.

Ateyah, A. H. (2022). Bending stresses calculation in welded joint. *International Journal of Engineering Research and Applications, 12*(1), 1–6. https://doi.org/10.9790/9622–1201010106.

Atmaja, A. P., Hakim, A. E., Wibowo, A. A., & Pratama, L. A. (2021). Communication systems of smart agriculture. *Journal of Robotics and Control, 2*(4), 297–301. https://doi.org/10.18196/jrc.2495.

Axiomatic Technologies Corporation. (2018). *Technical datasheet #TDAX141100.* Retrieved May 15, 2021, from Axiomatic Global Electronic Solutions. https://www.axiomatic.com/TDAX141100.pdf

Axiomatic Technologies Corporation. (2020, July 23). *Technical datasheet #TDAX140900.* Retrieved May 15, 2021, from Axiomatic Global Electronic Solutions. https://www.axiomatic.com/TDAX140900.pdf

Bacenetti, J., Lovarelli, D., Facchinetti, D., & Pessina, D. (2018, May 3). An environmental comparison of techniques to reduce pollutants emissions related to agricultural tractors. *Biosystems Engineering, 171*, 30–40. https://doi.org/10.1016/j.biosystemseng.2018.04.014.

Banno, Y., & Kinoshita, K. (2022). Experimental investigation of fatigue strength of out-of-plane gusset welded joints under variable amplitude plate bending loading in long life region. *Welding in the World, 66*, 1883–1896. https://doi.org/10.1007/s40194-022-01312-6.

Batey, T. (2009). Soil compaction and soil management—a review. *Soil Use and Management, 25*(4), 335–345. https://doi:10.1111/j.1475-2743.2009.00236.x.

Behringer, M., & Kuehne, M. (2011). *Network complexity and how to deal with it*. Retrieved December 22, 2020, from Ripe NCC. https://labs.ripe.net/Members/ mbehring/network-complexity-and-how-to-deal-with-it

Bjork, T., Ahola, A., & Skriko, T. (2020). On the distortion and warping of cantilever beams with hollow section. *Welding in the World, 64*, 1269–1278. https://doi.org/10.1007/s40194-020-00911-5.

Blodgett, O. W. (1972). *Design of welded structures*. The James F. Lincoln Arc Welding Foundation.

Botta, G. F., Tolon-Becerra, A., Bienvenido, F., Rivero, D., Laureda, D. A., Ezquerra-Canalejo, A., & Contessotto, E. E. (2018). Sunflower harvest: Tractor and grain chaser traffic effects on soil compaction and crop yields. *Land Degradation & Development, 29*, 4252–4261. https://doi.org/10.1002/ldr.3181.

Botta, G. F., Tolon-Becerra, A., Tourn, M., Lastra-Bravo, X., & Rivero, D. (2012). Agricultural traffic: Motion resistance and soil compaction in relation to tractor design and different soil conditions. *Soil and Tillage Research, 120*, 92–98. https://doi.org/10.1016/j.still.2011.11.008.

Brehmer, A. (2014). *CAN based protocols in avionics*. Retrieved May 15, 2021, from Vector. https://ieeexplore.ieee.org/stamp /stamp.jsp?arnumber=6979561

Brown, T. H., J. R. (2005). *Marks' calculations for machine design*. McGraw-Hill. https://doi.org/10.1036/0071466916.

Budynas, R., & Nisbett, K. (2020). *Shigley's mechanical engineering design* (11th ed.). McGraw Hill. ISBN: 978–0073398211Burris, D. L., & Sawyer, W. G. (2006). A low friction and ultra low wear rate PEEK/PTFE composite. *Wear, 261*(3–4), 410–418. https://doi.org/10.1016/j.wear.2005.12.016.

Bushing MFG. (2023). *PTFE bearings*. Retrieved December 27, 2023, from https://bushingmfg.com/ptfe-bearings/

Calonego, J. C., Raphael, J. P., Rigon, J. P., de Olieria-Neto, L., & Rosolem, C. A. (2017). Soil compaction management and soybean yields with cover crops under no-till and occasional chiseling. *European Journal of Agronomy, 85*, 31–37. https://doi.org/10.1016/j.eja.2017.02.001.

Canadian Welding Bureau. (2023). *Electrode certifications*. Retrieved May 1, 2023, from CWB Group. https://www .cwbgroup.org/safety/industry/electrode-certification

Carrera, A. (2023). *How do self-lubricating bearings lubricate?* Retrieved December 27, 2023, from tstar.com. https://www. tstar.com/blog/qa-how-do-self-lubricating-bearings-lubricate

Cary, H. B. (2002). *Modern welding technology*. Prentice Hall. ISBN: 978-1-68584-571-1

CASE IH. (2020a). *AFS*. Retrieved May 15, 2021, from CASE IH Agriculture. https://www.caseih.com/apac/en-int /products/advanced-farming-system

CASE IH. (2020b). *AFS connect*. Retrieved May 15, 2021, from CASE IH Agriculture. https://www.caseih.com/apac /en-int/products/advanced-farming-system/afs-connect

Case IH. (2021). *Steiger & Quadtrac*. CNH Industrial America, LLC. Retrieved November 23, 2021, from https://www. caseih.com/emea/en-za/products/tractors/steiger-quadtrac-series/steiger-quadtrac

Caterpillar. (2020). *Telematics—Your link to equipment cost savings*. Retrieved May 15, 2021, from CAT. https://www.cat .com/en_US/articles/customer-stories/forestry/telematics.html

Caterpillar. (2021). *1900s*. (Caterpillar, Inc.) Retrieved November 26, 2021, from Caterpillar History. https://www.cater-pillar.com/en/company/history/1900.html

Chamen, T. (2015). Controlled traffic farming—From worldwide research to adoption in Europe and its future prospects. *Acta Technologica Agrriculturae, 18*(3), 64–73. https://doi.org/10.1515/ata-2015–0014.

Chhabra, N. (2013). Comparative analysis of different wireless technologies. *International Journal of Scientific Research in Network Security and Communication, 1*(5), 13–17. https://doi: 2321–3256.

Chouha, E., Jalal, S. E., El Maskaoui, Z., & Chouaf, A. (2019). Concentrated stress location areas for welded tubular T-joint under deflected bending load. *MATEC Web of Conferences, 286*, 02004. https://doi.org/10.1051/matecconf /201928602004.

CiA. (2010). Isobus—the CAN-based network system. *CAN Newsletter*, 44–48. Retrieved May 15, 2021, from https:// www.scribd.com/document/479072939/10-1-p44-isobus-pdf

CiA. (2017). *CAN-based higher-layer protocols (HLP)*. Retrieved May 15, 2021, from CAN in Automation. https://www .can-cia.org/can-knowledge/hlp/higher-layer-protocols/

CiA. (2018). *History of CAN technology*. Retrieved May 15, 2021, from CAN in Automation. https://www.can-cia.org/can-knowledge/can/can-history/

Copperhill Technologies. (2020a). *A brief introduction to controller area network*. Retrieved May 15, 2021, from Copperhill Technologies. https://copperhilltech.com/a-brief-introduction-to-controller-area-network/

Copperhill Technologies. (2020b). *A brief introduction to the SAE J1939 protocol*. Retrieved May 15, 2021, from Copperhill Technologies. https://copperhilltech.com/a-brief-introduction-to-the-sae-j1939-protocol/

CSS Electronics. (2020). *CANopen explained—A simple intro*. Retrieved May 15, 2021, from https://www.csselectronics.com/screen/page/canopen-tutorial-simple-intro/language/en

CSS Electronics. (2020a). *CAN Bus explained—A simple intro*. Retrieved May 15, 2021, from CSS Electronics. https://www.csselectronics.com/screen/page/simple-intro-to-can-bus/language/en

CSS Electronics. (2020b). *J1939 explained—A simple intro*. Retrieved May 15, 2021, from CSS Electronics. https://www.csselectronics.com/screen/page/simple-intro-j1939-explained

CSS Electronics. (2020c). *OBD2 explained—A simple intro*. Retrieved May 15, 2021, from CSS Electronics. https://www.csselectronics.com/screen/page/simple-intro-obd2-explained

Čupera, J., & Sedlák, P. (2011). The use of CAN-Bus messages of an agricultural tractor for monitoring its operation. *Research in Agricultural Engineering, 57*(4), 117–127. https://doi.org/10.17221/20/2011-RAE.

Darr, M. J. (2012). CAN Bus technology enables advanced machinery management. *Resource Magazine, 19*, no. 5 (2012): 10–11.

Davies, D. B., Finney, J. B., & Richardson, S. J. (1973). Relative effects of tractor weight and wheel-slip in causing soil compaction. *Journal of Soil Science, 24*, 399–409. https://doi.org/10.1111/j.1365-2389.1973.tb00775.x.

Dawe, J. L., & Kulak, G. L. (1974). Welded connections under combined shear and moment. *Journal of the Structural Division, 100*(4), 727–741. https://doi.org/10.1061/JSDAEG.0003753.

Deutschman, A. D., Michels, W. J., & Wilson, C. E. (1975). *Machine design—Theory and practice*. Macmillan. ISBN: 978–0023290008

Dorle, S., Deshpande, D., Keskar, A., & Chakole, M. (2010). *Vehicle classification and communication using Zigbee protocol*. 2010 3rd International Conference on Emerging Trends in Engineering and Technology, 106–109. https://doi.org/10.1109/ICETET.2010.170.

Dozier, I. A., Behnke, G. D., Davis, A. S., Nafziger, E. D., & Villamil, M. B. (2017). Tillage and cover cropping effects on soil properties and crop production in Illinois. *Agronomy Journal, 109*, 1261–1270. https://doi.org/10.2134/agronj2016.10.0613.

Duiker, S. (2005). *Avoiding soil compaction*. Retrieved November 21, 2021, from PennState Extension. https://extension.psu.edu/effects-of-soil-compaction

Dwivedi, D. K. (2022). *Fundamentals of metal joining*. Springer. https://doi.org/10.1007/978-981-16-4819-9.

Elmer's Manufacturing, Inc. (2016). *The benefits of tracks vs. tires*. Retrieved November 24, 2021, from Elmer's Manufacturing. https://elmersmfg.com/2016/04/benefits-tracks-vs-tires/

Embitel. (2018). *Technology behind telematics explained: How does a vehicle telematics solution work?* Retrieved May 15, 2021, from Embitel: embitel.com/blog/embedded-blog/tech-behind-telematics-explained-how-does-a-vehicle-telematics-solution-work

Engineering Toolbox. (2008). *BHN—Brinell hardness number*. Retrieved December 18, 2021, from Engineering Toolbox. https://www.engineeringtoolbox.com/bhn-brinell-hardness-number-d_1365.html

Fanton, D. (2020, July 10). *Why all the fuss about CAN Bus?* Retrieved May 15, 2021, from I/O Hub. https://www.onlogic.com/company/io-hub/fuss-can-bus/

Fitch, B. (2023). *The grease gun: Applications, uses, and benefits*. Retrieved December 27, 2023, from Machinery Lubrication. https://www.machinerylubrication.com/Read/29356/grease-gun-anatomy

Forsthoffer, M. (2019). Journal (radial) bearings. In M. Forsthoffer, *Forsthoffer's component condition monitoring*. Butterworth-Heinemann. Retrieved December 27, 2023, from https://www.sciencedirect.com/topics/engineering/antifriction-bearings

Franklin Fibre. (2023). *Essential design practices for composite bearings and bushings*. Retrieved December 27, 2023, from Franklin Fibre. https://www.franklinfibre.com/blog/essential-design-practices-for-composite-bearings-

and-bushings#:~:text=For%20best%20design%2C%20the%20length,cause%20edge %20concentration%20of%20 loading.&text=%E2%80%8DThe%20smoother%20the%20surface%20of,16RMS%20or%20bet

Gao, X., Huang, D., Chen, Y., Jin, W., & Luo, Y. (2013). The design of a distributed control system based on CAN bus. *IEEE International Conference on Mechatronics and Automation*, 1118–1122. https://doi.org/10.1109/ICMA.2013.6618071.

Gary W. Krutz, J. K. (1994). *Machine design for mobile and industrial applications*. Society of Automotive Engineers, Inc. ISBN: 978-1560913894

GGB by Timken. (2019). *How does a self-lubricating bearing work?* Retrieved December 27, 2023, from GGP by Timken. https://www.ggbearings.com/en/why-choose-ggb/faq/bearings-faq/what-self-lubricating-bearing

GGB by Timken. (2020). *Composite bearing design with improved tribology*. GGB. Retrieved December 27, 2023, from https://www.ggbearings.com/sites/default/files /inline-files/GGB-Whitepaper-Composite-Bearings-for-Aggressiv e-Applications.pdf

GGB by Timken. (2023). *Fiber reinforced composite bearing handbook*. Retrieved December 27, 2023, from https://www. ggbearings.com/sites/default/files/2023-08/GGB-Fiber-Reinforced-Composite-Bearings-High-Load-Self-Lubricating-Bearings-Bushings_0.pdf

Ghosh, U., & Daigh, A. L. (2020). Soil compaction problems and subsoiling effects on potato crops: A review. *Crop, Forage and Turfgrass Management, 6*, 1–10. https://doi.org/10.1002/cft2.20030.

Gibson, R. F. (1994). *Principles of composite material mechanics*. McGraw-Hill. ISBN: 978-1498720694

Goering, C., Stone, M., Smith, D., & Turnquist, P. (2006). *Off-road vehicle engineering principles*. American Society of Agricultural Engineers. ISBN: 978-1892769268

Gomez, I., Kanvinde, A., Kwan, Y. K., & Grondin, G. (2008). *Strength and ductility of welded joints subjected to out-of-plane bending*. Retrieved May 25, 2024, from American Institute of Steel Construction. https://www.aisc .org/globalassets/aisc/research-library/strength-and-ductility-of-welded-joints-subjected-to-out-of-plane-bending.pdf

Gunjal, K., Lavoie, G., & Raghavan, G. S. (1987). Economics of soil compaction due to machinery traffic and implications for machinery selection. *Canadian Journal of Agricultural Economics/Revue canadienne d'agroeconomie, 35*, 591–603. https://doi.org/10.1111/j.1744-7976.1987.tb02251.x.

Hadjilambrinos, C. (2021). Reexamining the automobile's past: What were the critical factors that determined the emergence of the internal combustion engine as the dominant automotive technology? *Bulletin of Science, Technology, and Society, 41*(2–3), 58–71. https://doi.org/10.1177/02704676211036334.

Hawkins, E. M. (2015). Benchmarking costs of fixed-frame, articulated, and tracked tractors. *Applied Engineering in Agriculture, 31*(5), 741–745. https://doi.org/10.13031/aea.31.11074.

Hectors, K., & De Waele, W. (2021). Cumulative damage and life prediction models for high-cycle fatigue of metals: a review. *Metals, 204*(11). https://doi.org/10.3390/met11020204.

Hectors, K., & De Waele, W. (2021). Influence of weld geometry on stress concentration factor distribution in tubular joints. *Journal of Construction Steel Research, 176*, 106376. https://doi.org/10.1016/j.jcsr.2020.106376.

Henninger, F., & Friedrich, K. (2002). Thermoplastic filament winding with on-line impregnation: Part A. Process technology and operating efficiency. *Composites Part A: Applied Science and Manufacturing, 33*(11), 1479–1486. https://doi. org/10.1016/S1359-835X(02)00135-5.

Henninger, F., Hoffmann, J., & Friedrich, K. (2002). Thermoplastic filament winding with on-line impregnation: Part B. Experimental study of processing parameters. *Composites Part A: Applied Science and Manufacturing, 33*(12), 1684–1695. https://doi.org/10.1016/S1359-835X(02)00136-7.

Hobbacher, A. (2016). *International Institute of Welding: Recommendations for Fatigue Design of Welded Joints and Components*. Springer International Publishing. https://doi: 10.1007/978-3-319-23757-2.

Hobbacher, A., & Kassner, M. (2012). On the relations between fatigue properties of welded joints, quality criteria and groups in ISO 5817. *Welding in the World, 56*, 153–169. https://doi: 10.1007/BF03321405.

Horn, R., Domzzał, H., Słowińska-Jurkiewicz, A., & van Ouwerkerk, C. (1995). Soil compaction processes and their effects on the structure of arable soils and the environment. *Soil and Tillage Research, 35*, 23–36. https://doi. org/10.1016/0167-1987(95)00479-C.

Horton, M. (2019). *What can a CANbus IMU do to make an autonomous vehicle safer?* Retrieved May 15, 2021, from Autonomous Vehicle International. https://www.autonomousvehicleinternational.com/opinion/what-can-a-canbus-imu-do-to-make-an-autonomous-vehicle-safer.html

Hoseinian, S. H., Hemmat, A., Esehaghbeygi, A., Shahgoli, G., & Baghbanan, A. (2021). Development of a dual sideway-share subsurface tillage implement: Part 2. Effect of tool geometry on tillage forces and soil disturbance characteristics. *Soil and Tillage Research, 215,* 105200. https://doi.org/10.1016/j.still.2021.105200.

Husaini, N. A., Hamza, J. K., & Sofyan, S. E. (2019). Effects of welding on the change of microstructure and mechanical properties of low carbon steel. *Material Science and Engineering, 523,* 012065. https://doi.org/10.1088/1757–899X/523/1/012065.

Igus Motion Plastics. (2023). *Bearings are rigorously tested within the iglide® test laboratory.* Retrieved December 27, 2023, from Igus Motion Plastics. https://www.igus.com/info/plain-bearings-testlaboratory

Igus Motion Plastics. (2023). *iglide® Q: able to withstand high loads, even when subjected to the strain of tumbling.* Retrieved December 27, 2023, from Igus Motion Plastics. https://www.igus.com/info/plain-bearings-test-iglide-q

Igus Motion Plastics. (2023). *Igus engineer's toolbox plain bearing design guide.* Retrieved December 27, 2023, from Igus Plain Bearing Design Guide. https://toolbox.igus.com/design-guides/iglide-plain-bearings-design-guide

Industrial Training Partners Ltd. (1983). *Weld defects: Causes and Corrections.*

Institute of Electrical and Electronics Engineers. (2016). *IEEE 802.11-2016—IEEE Standard for Information technology—Telecommunications and information exchange between systems local and metropolitan area networks—Specific requirements—Part 11.*

Institute of Electrical and Electronics Engineers. (2020). *IEEE 802.15.4-2020—IEEE Standard for Low-Rate Wireless Networks.* Retrieved May 1, 2020, from https://ieeexplore.ieee.org/document/9144691

Islamovic, F. I., Muratovic, P., Gaco, D., & Kulenovic, F. (2009). *Bend testing of the welded joints.* 7th International Scientific Conference on Production Engineering (RIM 2009)—Cairo. Bihac, Bosnia and Herzegovina: Society for Robotics of Bosnia and Herzegovina.

ISO. (2006). *ISO 11898-3:2006.* Retrieved May 15, 2021, from International Organization for Standardization. https://www.iso.org/standard/36055.html

ISO. (2015). *ISO 11898-1:2015.* Retrieved May 15, 2021, from International Organization for Standardization. https://www.iso.org/standard/63648.html

ISO. (2016). *ISO 11898-2:2016.* Retrieved May 15, 2021, from International Organization of Standardization. https://www.iso.org/standard/67244.html

ISO. (2020). *About us.* Retrieved May 15, 2021, from International Organization for Standardization. https://www.iso.org/about-us.html

Jadhav, G. R., Zoman, D. B., & Mahajan, D. R. (2015). Analysis of composite journal bearing. *International Engineering Research Journal, 1998–2002.* Retrieved December 27, 2023, from https://www.ierjournal.org/pupload/mitpgcon/1998-2002.pdf

Jadhav, G., Badguiar, T., Zorman, D. B., & Jadhav, M. (2016). Tribological performance analysis of composite materials for journal bearing. *International Journal of Modern trends in Engineering and Research, 3.* https://doi.org/10.21884/ijmter.2016.3017.apwts.

Jahn Research Group. (2019). *Cyber risk and security implications in smart agriculture and food systems.* University of Wisconsin-Madison White Paper. Madison. Retrieved May, 15, 2021, from https://jahnresearchgroup.webhosting.cals.wisc.edu/wp-content/uploads/sites/223/2019/01/Agricultural-Cyber-Risk-and-Security.pdf

Jamail, H., Nachimuthy, G., Palmer, B., Hodgson, D., Hundt, A., Nunn, C., & Braunack, M. (2021). Soil compaction in a new light: Knowing the cost of doing nothing—a cotton case study. *Soil & Tillage Research, 213,* 105158. https://doi.org/10.1016/j.still.2021.105158.

Jasa, P. (2019). *Avoiding harvest compaction in wet soils.* University of Nebraska-Lincoln Institute of Agriculture and Natural Resources. Retrieved November 29, 2021, from Cropwatch. https://cropwatch.unl.edu/2019/avoiding-compaction-harvest

John Deere. (2020a). *Operations center.* Retrieved May 15, 2021, from Deere & Company. https://www.deere.com/en/technology-products/precision-ag-technology/data-management/operations-center

John Deere. (2020b). *Gen 4 command center display*. Retrieved May 15, 2021, from Deere & Company. http://www.deere.com/en/technology-products/precision-ag-technology/ guidance/gen-4-premium-activation/

Jonsson, B., Samuelsson, J., & Marquis, G. B. (2011). Development of weld quality criteria based on fatigue performance. *Welding in the World, 55*, 79–88. https://doi.org/10.1007/BF03321545.

June, M. (2014). *Mechanics of Materials-Steel*. Retrieved December 18, 2021, from Learn Civil Engineering.com. http://www.learncivilengineering.com/wp-content/themes/thesis/images/structural-engineering/Structural-steel-structural-light-gage-reinforcing.pdf

Kakani, S. L., & Kakani, A. (2012). *Material science*. New Age International Publishers, Ltd. ISBN: 978–1822426564

Kamilaris, A., Gao, F., Prenafeta-Boldu, F., & Ali, M. (2016). *A semantic framework for Internet of Things–enabled smart farming applications*. 2016 IEEE 3rd World Forum on Internet of Things, 442–447. https://doi.org/10.1109/WF-IoT.2016.7845467.

Karandikar, P. M., Kharde, R. R., Bhoyar, S. B., & Kadu, R. L. (2014). Study the tribological properties of PEEK/PTFE reinforced with glass fibers and solid lubricants at room temperature. *International Journal of Current Engineering and Technology, 4*(4), 2401–2404. Retrieved December 27, 2023, from https://inpressco.com/wp-content/uploads/2014/07/Paper192401-2404.pdf

Kassel, S. (1971). *Lunokhod-1 Soviet lunar surface*. RAND Corporation. Retrieved May 15, 2022 from https://www.rand.org/pubs/reports /R0802.html

Keller, T., & Arvidsson, J. (2004). Technical solutions to reduce the risk of subsoil compaction: Effects of dual wheels, tandem wheels and tyre inflation pressure on stress propagation in soil. *Soil and Tillage Research, 79*, 191–205. https://doi.org/10.1016/j.still.2004.07.008.

Kim, S.-H., Lee, C.-H., Ryu, I., & Park, S. (2023). Experimental investigation of cold formed high strength steel tubular joints subjected to moment loading. *Advances in Structural Engineering, 26*(12), 2172–2198. https://doi.org/10.1177/13694332221149501.

Kraus, D., Leitgeb, E., Plank, T., & Löschnigg, M. (2016). *Replacement of the controller area network (CAN) protocol for future automotive bus system solutions by substitution via optical networks*. 18th International Conference on Transparent Optical Networks (ICTON), 1–8. https://doi.org/10.1109/ICTON.2016.7550335.

Kvaser. (2017). *CAN Bus error handling*. Retrieved May 15, 2021, from Kvaser. https://www.kvaser.com/about-can /the-can-protocol/can-error-handling/

Kvaser. (2020). *The Kvaser dictionary for help in understanding CAN Bus systems*. Retrieved May 15, 2021, from Kvaser. https://www.kvaser.com/about-can/can-dictionary/

Kvaser. (n.d.). *Higher layer protocols*. Retrieved May 15, 2021, from https://www.kvaser.com/about-can /higher-layer-protocols/

Lamandé, M., Greve, M. H., & Schjønning, P. (2018). Risk assessment of soil compaction in Europe—Rubber tracks or wheels on machinery. *Catena, 167*, 353–362. https://doi.org/10.1016/j.catena.2018.05.015.

Lane, M. (1980). *The story of the steam plough works : Fowlers of Leeds*. Northgate Publishing Co. https://doi.org/10.0852984146.

Lee, J.-S., Su, Y.-W., & Shen, C.-C. (2007). *A comparative study of wireless protocols: Bluetooth, UWB, ZigBee, and Wi-Fi*. IECON 2007: 33rd Annual Conference of the IEEE Industrial Electronics Society. https://doi.org/10.1109/IECON.2007.4460126.

Li, Y., Yu, C., & Li, H. (2010). *The design of ZigBee communication convertor based on CAN*. 2010 International Conference on Computer Application and System Modeling (ICCASM 2010). https://doi.org/10.1109/ICCASM.2010.5620516.

Liang, J., Liu, X., & Liao, K. (2018). Soil moisture retrieval using UWB echoes via fuzzy logic and machine learning. *EEE Internet of Things Journal, 5*(5), 3344–3352. https://doi: 10.1109/JIOT.2017.2760338.

Lincoln Electric Company. (2000). *The procedure handbook of arc welding*. Lincoln Electric Company. ISBN: 978–9990022964

Lincoln Global, Inc. (2015). *Parts of a weld poster (WC-482)*. Retrieved December 18, 2021, from Lincoln Arc Welding. https://www.lincolnelectric.com/assets/US/EN /literature/WC482.pdf

Lindgren, M., & Hansson, P.-A. (2002). PM-power and machinery: Effects of engine control strategies and transmission characteristics on the exhaust gas emissions from an agricultural tractor. *Biosystems Engineering, 83*(1), 55–65. https://doi.org/10.1006/bioe.2002.0099.

Lovarelli, D., & Bacenetti, J. (2017). Bridging the gap between reliable data collection and environmental impact for mechanized field operations. *Biosystems Engineering, 160*, 109–123. https://doi.org/10.1016/j.biosystemseng.2017.06.002.

Lyasko, M. (2010). Slip sinkage effect in soil–vehicle mechanics. *Journal of Terramechanics, 47*, 21–31. https://doi.org/10.1016/j.jterra.2009.08.005.

Maradana, S. (2012). *CAN basics.* Retrieved May 15, 2021, from Automotive Techies. https://automotivetechis.wordpress.com/2012/06/01/can-basics-faq/

Mayr, C. M., & Rother, K. (2019). Improved stress concentration factors for circular shafts for uniaxial and combined loading. *Materials Testing, 61*(3), 193–203. https://doi.org/10.3139/120.111305.

McPheron, T. J., & Stwalley, R. M., III. (2022). Engineering challenges associated with welding filed repairs. In K. Cooke, *Welding Principles and Application.* IntechOpen Limited. https://doi.org/10.5772/intechopen.104263.

McPheron, T. J., & Stwalley, R. M., III. (2023). *Fillet weld strength analysis for cantilever loading.* ASABE 2023 AIM—Omaha Conference, St. Joesph, Michigan. https://doi.org/10.13031/aim.202300013.

Mellgren, P. G. (1980). Terrain classification for Canadian forest. *Canadian Pulp and Paper Association Journal*, 1–13.

Merchant, S. Y. (2015). Investigation on effect of welding current on welding speed and hardness of haz and weld metal of mild steel. *International Journal of Research in Engineering and Technology, 4*(3), 44–48. ISSN: 2319–1163

MetalTekInternational. (2020). *How to evaluate materials.* Retrieved December 18, 2021, from Metaltek Blog. https://www.metaltek.com/blog/how-to-elevate-materials-properties-to-consider/

Miller Electric Manufacturing, LLC. (2007). *Deciphering weld symbols.* Retrieved December 18, 2021, from Miller Electric. https://www.millerwelds.com/resources/article-library/deciphering-weld-symbols

Miller Electric Manufacturing, LLC. (n.d.). *Multimatic® 255 multiprocess welder w/EZ-Latch™ dual cylinder running gear & TIG kit—208–575V.* Retrieved May 15, 2024, from MillerWelds.com. https://www.millerwelds.com/equipment/welders/mig-gmaw/multimatic-255-multiprocess-welder-m30175

Misser Uitgeverij B.V. (2021). *NEXAT redesigning versatility, reducing field traffic.* Retrieved January 6, 2022, from Future Farming. https://www.futurefarming.com/tech-in-focus/autonomous-semiauto-steering/autonomous-vehicles/nexat-redesigning-versatility-reducing-field-traffic/

Moinfar, A., Shahgholi, G., Abbaspour-Gilandeh, Y., Herrera-Miranda, I., Hernandez-Hernandez, J. L., & Herrera-Miranda, M. A. (2021). Investigating the effect of the tractor drive system type on soil behavior under tractor tires. *Agronomy, 11*(4). https://doi.org/10.3390/agronomy11040696.

Morabito, R., Cozzolino, V., Ding, A., Beijar, N., & Ott, J. (2018). Consolidate IoT edge computing with lightweight virtualization. *IEEE Network, 32*(1), 102–111. https://doi.org/10.1109/mnet.2018.1700175.

Moran, C. D. (2021). *Interpreting metal fab drawings.* Open Oregon Educational Resources. Retrieved December 18, 2021, from https://openoregon.pressbooks.pub/weldsymbols/chapter/interpreting-metal-fab-drawings-2/

Munkholm, L. J., & Schjonning, P. (2004). Structural vulnerability of a sandy loam exposed to intensive tillage and traffic in wet conditions. *Soil and Tillage Research, 79*, 79–85. https://doi.org/10.1016/j.still.2004.03.012.

Mupoperi, T. R., & Pita, M. (2022). *An investigation of the effect of welding current on the mechanical properties of mild steel joints when using arc welding.* IEEE 13th International Conference on Mechanical and Intelligent Manufacturing Technologies—Cape Town. IEEE. https://doi.org/10.1109/ICMIMT55556.2022.9845346.

National Board of Boiler and Pressure Vessel Inspectors. (2014). *Identifying existing materials.* Retrieved December 18, 2021, from 2014-8_TechPresentation_Beach_Scribner. https://www.nationalboard.org/SiteDocuments/Members%20Only/Technical%20Presentations/2014-8_TechPresentation_Beach_Scribner.pdf

National Instruments Corp. (2020). *FlexRay automotive communication bus overview.* Retrieved from National Instruments Corp. May 8, 2021, from https://www.ni.com/en-us/innovations/white-papers/06/flexray-automotive-communication-bus-overview.html

National Inventors Hall of Fame. (2006). *Benjamin Holt track-type tractor.* Retrieved November 30, 2021, from National Inventors Hall of Fame Inductees. https://www.invent.org/inductees/benjamin-holt

National Lubricating Grease Institute. (2017). *Grease glossary.* Retrieved December 27, 2023, from National Lubricating Grease Institute. https://www.nlgi.org/grease-glossary/nlgi-grade/

Navixy. (2020). *CAN bus and alternatives.* Retrieved May 15, 2021, from Navixy. https://www.navixy.com/docs/academy/can-bus/can-and-alternatives/

Nawaz, M. F., Bourrie, G., & Trolard, F. (2013). Soil compaction impact and modelling. A review. *Agronomy for Sustainable Development, 33*, 291–309. https://doi.org/10.1007/s13593-011-0071-8.

Nguyen, O. (2018). *3 most common industries for MIG welding.* Retrieved December 18, 2021, from Tulsa Welding School. https://www.tws.edu/blog/welding/3-most-common-industries-for-mig-welding/

Nikander, J., Manninen, O., & Laajalahti, M. (2020). Requirements for cybersecurity in agricultural communication networks. *Computers and Electronics in Agriculture, 179*, 1–10. https://doi.org/10.1016/j.compag.2020.105776.

Norman, P. (2021). *IF and VF tyres—What are they?* Retrieved November 21, 2021, from Brocks Wheel & Tyre. https://bwt.uk.com/news/if-tyres-and-vf-tyres-what-are-they/

NTS Tire Supply Team. (2019). *Tires vs. tracks: Which creates less compaction?* Retrieved December 1, 2021, from Practical Traction Knowledge. https://www.ntstiresupply.com/ptk-shared/tires-vs-tracks-which-creates-less-compaction

Ohio Department of Transportation. (2011). *Field welding inspection guide.* Retrieved December 18, 2021, from Office of Materials Management. https://www.dot.state.oh.us/Divisions/ConstructionMgt/Materials/Miscellaneous/Field-Welding-Inspection-Guide.pdf

Pathak, D., Singh, R. P., Gaur, S., & Balu, V. (2020). Experimental investigation of effects of welding current and electrode angle on tensile strength of shielded metal arc welded low carbon steel plates. *Materials Today: Proceedings, 26*, 929–931. https://doi.org/10.1016/j.matpr.2020.01.146.

Phillips, D. H. (2016). *Welding engineering: An introduction.* John Wiley & Sons. ISBN: 978-1-119-85872-0

Pilkey, W. D., & Pilkey, D. F. (2008). *Peterson's stress concentration factors.* John Wiley & Sons. https://doi.org/10.1002/9780470211106.

Polcar, A., Čupera, J., & Kumbár, V. (2016). Calibration and its use in measuring fuel consumption with the CAN-Bus network. *Acta Universitatis Agriculturaevet Silviculturae Mendelianae Brunensis, 64*(2), 503–507. https://doi.org/10.11118/actaun201664020503.

Polygon Composites Technology. (2023). *Bearings design guide.* Retrieved December 27, 2023, from https://polygoncomposites.com/design-guide-bearings/

Polygon Composites Technology. (2023). *Case studies.* Retrieved December 27, 2023, from Polygon Composites Technology. https://polygoncomposites.com/case-studies/

Polygon Composites Technology. (2023). *Industries.* Retrieved December 27, 2023, from Polygon Composites Technology. https://polygoncomposites.com/industries/

Polygon Composites Technology. (2023). *Polygon and Kuhne Industrie delivers composite bearings for demanding mixer cart application.* Retrieved December 27, 2023, from Polygon Composites Technology. https://polygoncomposites.com/resources/kuhne-industrie-delivers-composite-bearings-mixer-cart-application/

Polygon Composites Technology. (2023). *Polygon composite bushings ensure reduced maintenance in bulldozer designs.* Retrieved December 27, 2023, from Polygon Composites Technology. https://polygoncomposites.com/resources/composite-bushings-ensure-reduced-maintenance-in-bulldozer-designs/

Polygon Composites Technology. (2023). *Polygon composite bushings simplify maintenance in K-Tec heavy-duty earthmoving equipment.* Retrieved December 27, 2023, from Polygon Composites Technology. https://polygoncomposites.com/resources/polygon-composite-bushings-simplify-maintenance-in-heavy-duty-equipment/

Prayogo, G., Budiyanto, M. A., & Perkasa, M. (2023). Analysis of bending mechanical performance of welding joints with the addition of diamond and circular plates. *Indonesian Journal of Engineering and Science, 4*(1), 57–73. https://doi.org/10.51630/ijes.v3.i1.94.

Qiu-dong, W., Yang, W., Zhong-qiu, F., & Yi-xun, W. (2023). Parametric study on fatigue failure modes of the rib-deck weld under out-of-plane bending loading. *Iranian Journal of Science and Technology, Transactions of Civil Engineering, 47*, 2625–2637. https://doi.org/10.1007/s40996-023-01080-3.

Quine, A. (2008, January 29). *Carrier sense multiple access collision detect (CSMA/CD) explained.* Retrieved May 8, 2021, from IT Professional's Resource Center. https://www.itprc.com/carrier-sense-multiple-access-collision-detect-csmacd-explained/

Raney, W. A., Edminster, T. W., & Allaway, W. H. (1955). Current status of research in soil compaction. *Soil Science Society of America Journal, 19*, 423–428. https://doi.org/10.2136/sssaj1955.03615995001900040008x.

Raper, R. L., Reeves, D. W., Burmester, C. H., & Schwab, E. B. (2000). Tillage depth, tillage timing, and cover crop effects on cotton yield, soil strength, and tillage energy requirements. *Applied Engineering in Agriculture, 16*(4), 379–385. https://doi.org/10.13031/2013.5363.

Rohrer, R., Pitla, S., & Luck, J. (2019). Tractor CAN bus interface tools and application development for real-time data analysis. *Computers and Electronics in Agriculture, 163*. https://doi.org/10.1016/j.compag.2019.06.002.

Santacruz, A., & Millelsen, O. (2021). *Numerical stress analysis of tubular joints*. Third Conference of Computational Methods & Ocean Technology—Stavanger. IOP Science. https://doi.org/10.1088/1757–899X/1201/1/012032.

saVRee. (2023). *Plain bearing*. Retrieved December 27, 2023, from saVRee. https://savree.com/en/encyclopedia /plain-bearing

Schiller, R., Oswald, M., Neuhausler, J., Rother, K., & Engelhardt, I. (2022). Fatigue strength of partial penetration butt welds of mild steel. *Welding in the World, 66*, 2563–2584. https://doi.org/10.1007/s40194-022-01335-z.

Sengsri, P., Marsico, M. R., & Kaewunruen, S. (2019). *Base isolation fibre-reinforced composite bearings using recycled rubber*. IOP Conference Series: Materials Science and Engineering 603. Prague. https://doi.org/10.1088/1757–899X/603/2 /022060.

SEU. (2017). *Welding myths debunked—Maximum fillet weld thickness*. Retrieved from Structural Engineering University. https://learnwithseu.com/welding-myths-debunked-maximum-fillet-weld-thickness/

Shah, A. N., Tanveer, M., Shahzad, B., Yang, G., Fahad, S., Ali, S., . . . Souliyanonh, B. (2017). Soil compaction effects on soil health and crop productivity: An overview. *Environmental Science and Pollution Research, 24*, 10056–10067. https:// doi.org/10.1007/s11356-017-8421-y.

Shaheb, M. R., Venkatesh, R., & Shearer, S. A. (2021). A review of the effect of soil compaction and its management for sustainable crop production. *Journal of Biosystems Engineering, 46*, 417–439. https://doi.org/10.1007/s42853-021-00117-7.

Sharda, A. (2020). *Is swarm farming the future of farming?* Retrieved May 15, 2021, from Precision Ag Reviews. https:// www.precisionagreviews.com/post/is-swarm-farming-the-future-of-farming

Shawver, C., Brummer, J., Ippolito, J., Ahola, J., & Rhoades, R. (2020). *Managing cattle impacts when grazing on wet soils*. Retrieved November 28, 2021, from Colorado State University Extension. https://extension.colostate.edu/topic-areas /agriculture/managing-cattle-impacts-when-grazing-on-wet-soils-1-634/

Silicon Labs, Inc. (n.d.). *ZigBee-based home area networks enable smarter energy management*. https://www.silabs.com /documents/public/white-papers/ZigBee-based-HANs-for-Energy-Management.pdf

SKF. (2012). *SKF composite plain bearings*. Retrieved December 27, 2023, from SKF. https://cdn.skfmediahub.skf.com /api/public/0901d19680229dfc/pdf_preview_medium/0901d19680229dfc_pdf_preview_medium.pdf

Smith, C. W., Johnston, M. A., & Lorentz, S. (1997). Assessing the compaction susceptibility of South African forestry soils: I. The effect of soil type, water content and applied pressure on uni-axial compaction. *Soil and Tillage Research, 41*, 53–73. https://doi.org/10.1016/S0167–1987(96)01084-7.

Soane, B. D., & van Ouwerkerk, C. (1994). Soil compaction problems in world agriculture. In *Developments in Agricultural Engineering* (Vol. 11, pp. 1–21). Elsevier B.V. https://doi.org/10.1016/B978-0-444-88286-8.50009-X.

Soane, B. D., & van Ouwerkerk, C. (1995). Implications of soil compaction in crop production for the quality of the environment. *Soil and Tillage Research, 35*, 5–22. https://doi.org/10.1016/0167–1987(95)00475-8.

Sontowski, S., Gupta, M., Chukkapalli, S. S., Abdelsalam, M., Mittal, S., Joshi, A., & Sandhu, R. (2020). *Cyber attacks on smart farming infrastructure*. UMBC. Retrieved May 15, 2021, from https://ebiquity.umbc.edu/paper/html/id/944 / Cyber-Attacks-on-Smart-Farming-Infrastructure

Spence, C. C. (1960). *God speed the plow: The coming of steam cultivation to Great Britain*. University of Illinois Press. https:// doi.org/10.2307/3101425.

Spoor, G. (2006). Alleviation of soil compaction: Requirements, equipment and techniques. *Soil Use and Management, 22*, 113–122. https://doi.org/10.1111/j.1475-2743.2006.00015.x.

Stone, M. L., McKee, K. D., Formwalt, C. W., & Benneweis, R. K. (1999). *An electronic communications protocol for agricultural equipment*. ISO 11783: An Electronic Communications Protocol for Agricultural Equipment: ASAE Distinguished Lecture #23, Agricultural Equipment Technology Conference, February 7–10 1999, Louisville, Kentucky, 1–17. https://doi: 913C1798.

T. R. Higgins, F. P. (1969). Proposed working stresses for fillet welds in building construction. *Engineering Journal, 6*, 16–20. https://www.aisc.org/Proposed-Working-Stresses-for-Fillet-Welds-in-Building-Construction.

Talabi, S. I., Owolabi, O. B., Adebisi, J. A., & Yahaya, T. (2014). Effect of welding variables on mechanical properties of low carbon steel welded joint. *Advances in Production Engineering & Management, 9*(4), 181–186. https://doi.org/10.14743/apem2014.4.186.

Tewari, S. P., Gupta, A., & Prakash, J. (2010). Effect of welding parameters on the weldability of material. *International Journal of Engineering Science and Technology, 2*(4), 512–516. ISSN: 0975-5462.

Titan International, Inc. (2014). *LSW technology.* Retrieved December 2, 2021, from Titan. https://www.titan-intl.com/innovation/lsw-tires

Tobe, Y., & Lawrence, F. V., Jr. (1977). Effect of inadequate joint penetration on fatigue resistance of high-strength structural steel welds. *Supplement to the Welding Journal, 9*, 259–266. OSTI: 6199368

Topcon. (2019). *Technical data sheet OPUS A3 ECO Full.* Retrieved May 15, 2021, from Montronica. http://www.motronica.com/wp-content/uploads/2019/08/TDS_A3e_FULL.pdf

Torbert, H. A., & Wood, C. W. (1992). Effects of soil compaction and water-filled pore space on soil microbial activity and N losses. *Communications in Soil Science and Plant Analysis, 23*, 1321–1331. https://doi.org/10.1080/00103629209368668.

TractorData. (2016). *Challenger 65.* Retrieved November 29, 2021, from https://www.tractordata.com/farm-tractors/000/9/1/917-challenger-65.html

Tulsa Welding School. (2020). *What are the different welding positions?* Retrieved December 18, 2021, from Welding. https://www.tws.edu/blog/welding/what-are-the-different-welding-positions/

Tuohy, S., Glavin, M., Hughes, C., Jones, E., Trivedi, M., & Kilmartin, L. (2014). Intra-vehicle networks: A review. *IEEE Transactions of Intelligent Transportation Systems*, 1–12. https://doi.org/10.1109/TITS.2014.2320605.

Tuschner, J. (2020). *Compare & contrast—Making the case for tires vs. tracks.* Retrieved November 30, 2021, from Farm Equipment. https://www.farm-equipment.com/articles/18193

University of Illinois. (1995). *Metals.* Retrieved December 18, 2021, from Materials Science & Technology (MAST). http://matse1.matse.illinois.edu/metals/metals.html

Vantsevich, V. V., & Blundell, M. V. (Eds.). (2015). *Advanced autonomous vehicle design for severe environments.* IOS Press. ISSN: 1874–6268

Voorhees, W. B. (1986). The effect of soil compaction on crop yield. *SAE Transactions, 95*(3), 1078–1084. Retrieved March 8, 2022, from https://www.jstor.org/stable/44725467

Voss, W. (2011). *A comprehensible guide to controller area network.* Copperhill Technologies Corporation. ISBN: 9780976511601

Voss, W. (2018, November 9). *Guide to SAE J1939—J1939 message format.* Retrieved May 15, 2021, from Copperhill Technologies. https://copperhilltech.com/blog/guide-to-sae-j1939-j1939-message-format/

Voss, W. (2019). *SAE J1939 bandwidth, busload and message frame frequency.* Retrieved May 15, 2021, from Copperhill Technologies. https://copperhilltech.com/blog/sae-j1939-bandwidth-busload-and-message-frame-frequency/

Wallheimer, B. (2019). *Purdue partnering on 5G research to improve ag automation.* Retrieved May 15, 2021, from Purdue University College of Agriculture. https://ag.purdue.edu/stories/purdue-partnering-on-5g-research-to-improve-ag-automation/

Wikipedia. (2021). Perseverance (rover). Retrieved November 26, 2021, from Wikipedia. https://en.wikipedia.org/wiki/Perseverance_(rover)

Williams, S. M., & Weil, R. R. (2004). Crop cover root channels may alleviate soil compaction effects on soybean crop. *Soil Science Society of America Journal, 68*, 1403–1409. https://doi.org/10.2136/sssaj2004.1403.

Winsor, S. (2012). *Healthy soil and profits from low-till.* Retrieved November 23, 2021, from Corn+Soybean Digest. https://www.farmprogress.com/tillage/healthy-soil-and-profits-low-till

Yang, X., Shu, L., Chen, J., Ferrag, M. A., Wu, J., Nurellari, E., & Huang, K. (2021). A survey on smart agriculture: Development modes, technologies, and security and privacy challenges. *IEEE/CAA Journal of Automatica Sinica, 8*(2), 273–302. https://doi.org/10.1109/JAS.2020.1003536.

Yu, H. N., Kim, S. S., & Lee, D. G. (2009). Optimum design of aramid-phenolic/glass-phenolic composite journal bearings. *Composites Part A: Applied Science and Manufacturing, 40*(8), 1186–1191. https://doi.org/10.1016/j.compositesa.2009.05.001.

Zabrodskyi, A., Sarauski, E., Kukharets, S., Juostas, A., Vasiliauskas, G., & Andriusis, A. (2021). Analysis of the impact of soil compaction on the environment and agricultural economic losses in Lithuania and Ukraine. *Sustainability, 13*, 7762. https://doi.org/10.3390/su13477762.

Zachary, L. W., & Burger, C. P. (1976). Stress concentrations in double welded partial joint penetration butt welds. *Welding Research Supplement, 55*(3), 77–82. Retrieved May 25, 2024, from http://worldcat.org/issn/00432296

Zuheir Barsoum, M. K. (2017). Ultimate strength capacity of welded joints in high strength steels. *Procedia Structural Integrity, 5*, 1401–1408. https://doi.org/10.1016/j.prostr.2017.07.204.

ABOUT THE AUTHORS

ROBERT M. STWALLEY III is an associate clinical professor at Purdue and teaches crop production equipment, power units and power trains, the design of off-road vehicles, computer numerical control (CNC) machining, and machine design in the Agricultural and Biological Engineering Department.

ROGER TORMOEHLEN is a professor in the Agricultural and Biological Engineering Department and the former head of the Department of Agricultural Sciences Education and Communication. He teaches safety in agriculture in the Agricultural and Biological Engineering program and is a highly regarded agricultural sciences educator and researcher.

www.ingramcontent.com/pod-product-compliance
Lightning Source LLC
Chambersburg PA
CBHW061420210326

41598CB00035B/6281